T0367558

The Men and the Sane

The Men and the Sane

Embaye Melekin

Library of Congress Control Number: 2015914327

ISBN:	Hardcover	978-1-5144-6173-0
	Softcover	978-1-5144-6172-3
	eBook	978-1-5144-6171-6

Print information available on the last page.

Rev. date: 09/07/2015

To order additional copies of this book, contact:
Xlibris
800-056-3182
www.Xlibrispublishing.co.uk
Orders@Xlibrispublishing.co.uk
720237

The history of the Jews and the history of Adolf Hitler as never told before. Some scientific facinations and end time gathering of the true Israelites.

http://www.abyssiniashallrise.com

In the year 730 BC the kingdom of Israel vanished into the thin air and only the kingdom of Judah remained. The kingdom of Judah was primarily composed of the tribes of Judah, Benjamin, Levi and a few numbers from each of the separated ten tribes of Israel that did not migrate to the north when their kinsmen separated and formed the northern kingdom of Israel. At the time there were half a million inhabitants in the kingdom of Judah. Fifty thousand were remnants of the house of Israel while the rest were mostly from the tribes of Judah, Benjamin and the Levites.

All the Israeli tribes lived together in harmony and the kingdom of Judah prospered greatly. Among them were thirty thousand members of the heathen nations who adored idols but lived under the dictates of the Israelites. The state was exclusively administered by Judah, Benjamin and the Levites and the other inhabitants had no role whatsoever in its governance. Hence, the remnants of the house of Israel and those from the heathen nation were primarily farmers and raised animals to sustain themselves. They had no influence in the affairs of the kingdom of Judah and played very little role in the commerce and economy of the nation.

The kingdom of Judah was ultimately invaded by the Babylonians and many of the creams of the nation carried away to Babylon. Over twenty-five thousand men from the tribes of Judah, Benjamin and Levi found themselves in Babylon. These tribes were targeted specifically by the Babylonians because they were the main engines running the kingdom of Judah. With nearly all the prominent men that mattered in the kingdom of Judah exiled to Babylon Nebuchadnezzar successfully crippled and reduced the people to submission and instituted his own dictates. Nebuchadnezzar's army burnt so many important buildings in Judah and transported every ornaments, utensils and records they could lay their hands on. After all the institutions in Judah crumbled and all the able bodied individuals that run them exiled to Babylon, the entire kingdom was in total shambles and only Babylonian dictates were observed by the people.

The exiled Israelites were dispersed around Babylon but majority were located in the capital city and their expertise exploited. The learned ones among them were placed in important positions and even in the king's palace and the artisans were made to work with very little pay. Soon, the Israelites began to accept their fate and made the best of their tragic situations.

They all married Babylonian women and were surprised that the women favored them because of their skills and their ability to earn a decent living. In a few years their numbers increased steadily and most of them were very happy in their new home. Nearly all integrated fully but a handful of them never forgot their origin and inculcated their faith and traditions upon their offspring. Many strived to teach their children the Aramaic language and couched them in the alphabets. However, not many were successful with the exception of a single family.

After ninety years of exile only Ezra's family spoke Aramaic and read in the Aramaic alphabets. Ezra's ancestor in exile was a very reluctant Babylonian and never accepted his ultimate fate. He ultimately had ten children from a Babylonian woman and he never spoke to his children in any other language but Aramaic. As a result, all his children spoke

Aramaic and understood the alphabets. Throughout his life he never learnt the Babylonian language and knew only a few words to get by. The tradition continued to his children but gradually subsequent generations of his began to fade into the Babylonian society and lost their ability to speak or write Aramaic. Ultimately, only Ezra's family would continue the wishes of their ancestor in exile. At the end only Ezra, his two sisters and his parents spoke and wrote Aramaic. No other Israeli in exile spoke or could read Aramaic.

Ezra was nineteen and was a fourth generation in exile. He was married and had two young babies. He was employed as a scribe in the king's palace because he was fluent and could write in both Farsi and Aramaic. One of his assignment as a scribe was to go through the king's archives and conduct researches and to understand past edicts imposed by previous rulers of Babylon. Hence, he spent most of his time reading numerous scrolls and binders containing the history of Babylon for the past thirteen hundred years. The archive was a huge building with hundreds of documents.

One day, when he was trying to locate the records of a particular king and the edicts that were passed throughout his reign, as ordered by the king, he unearthed a set of similar and huge binders that were buried under the piles of documents. When he opened the pages he discovered that they were written in Aramaic. When he read a few pages he knew that the books contained the history of the Israelites and he was very delighted. Most of the book had deteriorated over the years and most of them had their pages stuck together because of the load that was placed on them. And others the pages just crumpled into many pieces when he tried to turn a page. He counted twenty-nine bibles but only four were in good shape. And out of the four only one was almost brand new and in perfect shape. Ezra segregated the good bible and kept it in a unique location where he could easily locate it. Over the next few days he thought of various ways he could take out the bible from the palace and into his home. Ultimately, he stole the bible and got it out of the palace without being detected.

When the Babylonians invaded the kingdom of Judah almost a hundred years earlier they took all the bibles from the house of God and stored them in their archives as loots. Only one dilapidated bible was left in the entire kingdom of Judah after the exile. No scribes were left in Judah to make more copied. As a result, the only bible in Judah eventually crumbled and fell apart and there was no record of the Israelites to guide those that remained in Judah and they became ignorant of their history. Obviously, those in exile had no access to the bible and hence, became a lost people also.

The discovery and possession of the bible by Ezra will become very significant in the end. May years later the Babylonian empire would be invaded by a rival kingdom and the king's palace and all the documents in the archives would be burnt to ashes including the bibles. If Ezra had not taken out the bible it would have been the end of God's book and no one would have heard of it ever again. What are the odds that the only Israeli exile that spoke and read Aramaic, from among all the exiles, would be hired as a scribe in the king's palace and discover the bible and preserve it. Obviously, it was God's design to preserve his record.

Ezra spent his entire time reading the bible. He read the bible, cover to cover, two hundred times and almost memorized it. Ultimately, he would decide to translate the bible into the Babylonian language that the exiled Israelites spoke. It took him almost three years to translate the first copy and called it the Torah. Ezra occasionally encountered words he did not understand and he wrote the translations in a manner he assumed they meant. But his fault was when it came to the history of the northern kingdom of Israel. Since the kingdom of Israel had vanished permanently he felt no need to clog the Torah with their history. As a result, he skipped a total of thirty pages when translating the bible into the Torah, all the pages dealing with the northern kingdom of Israel. In a few instances also he replaced the name Israel by Judah and hence, the Torah became slightly different from the original book of the Israelites.

After Ezra translated the bible he began to make more and more copies. He enlisted and sought the aid of other Israeli scholars to assist him. In ten years there were a hundred copies of the Torah in Babylon. Ezra alone copied ten of them. Over that period of time, Ezra became a great preacher and organized the exiles into many groups each with its own synagogue and a copy of the Torah. Soon, the exiles understood their history and established a mode of worship based on the tenets of the Torah. Out of a population of a hundred and ninety exiles thirty thousand of them became ardent and fanatic followers of Ezra and rejected everything that was Babylonian. They saw themselves as the only chosen people of God and viewed the Babylonians as heathens and even an inferior set of people. Ezra worked so hard to recruit all the Israeli exiles but was unsuccessful. Most of the descendants of the exiles were immersed deep into the Babylonian culture that they did not see themselves as outsiders.

The devout Israelites became well organized and became a force to reckon with in Babylon. They obeyed their rabbis and walked strictly by the dictates of the Torah and began to observe Israeli traditions and cultures. Immediately Ezra told them that they were Jews and that they were the only blessed children of God from all the children of Israel. And they all sincerely believed him.

All the Jews segregated themselves from the other exiles and never welcomed them as brothers any longer. Eventually, Ezra told them that it was time for them to pack their belongings and return to Jerusalem. The problem was that the Babylonian king wouldn't let them leave his kingdom. When the Babylonian king inquired and demanded the reason why the Jews wanted to return to their homeland he was told that the Jews had a book and their God instructed them to return to their country. The king demanded to see the book and was read to him over many days. Ultimately, the king became a believer in the one God of Israel and vowed to help those who desired to return to the land of God, Israel. Thus, started the gradual migration of Jews to their original homeland. The king levied taxes to raise money to assist the returning Israelites. Eventually, and over a period of three years, more than thirty thousand Jews returned to Jerusalem and settled in the kingdom of Judah.

The Jews were already well organized in Babylon. In Judah, they set up their institutions quickly under the guidance of Ezra and took full control over the entire territories of Judah. They found some people that did not have any affinity with Israel or Judah and wondered why they occupied their land. They also discovered a people that claimed to be the remnants of the thirteen tribes of Israel. These set of people had no record of Israel but each knew their tribes and the origin of their ancestors. The Jews called them the lost tribes of Israel and discarded them as their inferiors.

Jews were prohibited from mixing their blood with the lost tribes of Israel by Ezra and subsequently by the rabbis that succeeded him. And no Jew did. The Jews became ruthless and uncompromising over the years and were dominant in every sphere of the state of Judah. A hundred years after their return from exile the Jews concluded to deport the heathen nations that had lived in Judah for many centuries. They encircled each heathen nation separately, and in different occasions, and lined them up and marched them to the closest boundary and drove them out of Judah. They were all sent to Syria. As a result, the inhabitants of Judah were only Jews and the lost tribes of Israel.

The Jews called their language Hebrew and it became the main language of communication in Judah. The Jews never attempted to learn Aramaic that the lost tribes of Israel spoke but it became a necessity for everyone to speak Hebrew. The Jews were exclusively all the priests, the judges and the lawmakers throughout Judah including among the lost tribes of Israel. Trade and commerce was also totally in the hands of the Jews and the lost tribes of Israel were relegated to farming and animal rearing only.

For a long time the Jews had cordial relationships with the lost tribe of Israel. As time went on, however, they began to distance themselves and eventually discarded them and equated them as nothing but heathens who belonged to non-Israeli nations. The Jews had anointed themselves as the only priests bestowed to preach Judaism and all the priests in the lost children of Israel's territories were Jews. But suddenly that changed. The Jews relegated them as non-Israelites and confiscated their Torah's and everyone worshipped in only Jewish synagogues. Only a single dilapidated Torah was in the midst of the lost tribes of Israel and that quickly wore out and discarded eventually. The lost tribes of Israel had no scribes to make more copies for the elites and educated people were exclusively Jews.

In around 110 BC the Jews notices a convoy of strange people riding on horsebacks. Their horses were unusually big and their bodies were covered with metal plates. The strangers were twenty in number and they roamed around Jerusalem and eventually settled at the outskirt of the town. They confiscated five goats from a nearby farm and built a huge fire. The next day they disappeared as they came and the Jews didn't know what to make of it. After twenty days, a huge army clothed in a similar manner as those they had seen earlier surrounded Jerusalem.

They were about two thousand in number. About five hundred were on horsebacks while the rest walked on foot. All of them were well armed. They didn't waste anytime and began to erect their tents at the outskirt of the town. Soon, there were ten different settlements surrounding Jerusalem. The people of Judah just watched silently wondering what the strangers were up to. As the evening approached a band of soldiers roamed around the villages and confiscated goats and sheep and dragged them to their camps.

The next day, a large man surrounded by a few soldiers covered with armored bodies and gathered the people and began to speak to them. He spoke in German with a commanding voice but the numerous people of Judah that had gathered understood not a word of what he said. The Roman general did not care if they understood his words or not. All that he knew was that he had declared Judah as a Roman territory and that was how Judah was occupied by Rome. And it didn't take the Romans a long time to organize the region and enforce their rules.

They quickly realized that there were two sets of people in the region. They knew that the Jews exclusively ruled Judah and that the rest of the population was their vassals. Hence, the Romans dealt primarily with the Jews and recognized their local authorities as long as they were loyal to Rome. As the years went by the Romans would chose about ten Jews and send them to Rome to study Roman laws, language and responsibilities. They returned to Judah and became the main intermediaries between the local people and the occupying power.

Soon, the Romans began to establish what they called "the people's militia." They began to pick able-bodied persons from the entire population of Judah but they encountered stiff oppositions from the Jews. No Jew was willing to accept any instructions enforced by the lost tribes of Israel. The Romans respected the Jewish concerns and the militias were composed only of Jews. At the end, the Romans will have about ten thousand militias. Members of the militia had Roman citizenship conferred upon them and declared as the defenders of the Roman empire and enjoyed special privileges different from the rest of the population. Only the tax collectors from the militias were paid and the rest only had different Roman titles conferred upon them. The militias became the eyes and ears of the Romans and enforced their rules with vigor and sometimes, brutality.

When the Romans initially conquered Judah the ordinary soldiers went on rampage and raped women, including married ones, and that brought uproar among the population. When the news was brought to the attention of the Roman governor he was appalled by the conducts of his soldiers. He decreed that no Roman soldier should have any sexual relationship with the local people. However, some of the soldiers sought consensual sexual favors from young women and there were a few who mated with them. Those that were suspected of having an affair with the Roman soldiers were virtually disowned by their parents and the people and no one married them. As a result, a few Roman bastards were born and when they grew up they were given special privileges because they had Roman blood. They normally recognized them by their hazel eyes and softer hair.

Thirteen years after the invasion of the Romans there was a severe drought in Judah. It lasted for two years. As a result, there was great shortage of food especially among the Jews. The lost tribes of Israel were mostly farmers and when the rains failed the first year they stored their grains and refused to sell them to the Jews. Out of the two hundred and ninety thousand Jews that inhabited Judah, ten thousand of them began to migrate in search of food. They all wanted to go to Syria, the closest region to them but were rejected by those they deported from Judah earlier and knew that they were not welcome. But still some of them took the risk and were not destroyed. Most of them, however, marched south and eastwards. The Egyptians closed their borders on them and the Jews roamed aimlessly among the various tribes that inhabited the Middle-east. Ultimately, some will settle in Lebanon,

Arabia and some even settle in as far land as Persia and Yemen. When the Romans saw the severity of the drought they consulted Rome and ships leaded with wheat were dispatched immediately. The inhabitants didn't get much but had enough for survival. Most of the Jews that migrated to Syria returned to Judah ultimately but the rest were never heard of when the drought finally subsided.

Then came Jesus Christ in around 30 AD. Prior to that Jesus was schooled among the Jews, preached in Jewish synagogues and advocated the messages of the Jewish prophets. The Jews accepted him as one of their own because his earthly father, Joseph was a Jew. Jesus was admired for his wisdom and powerful preaching in their synagogues and his special gift to heal all ailments and deformities. They Jews initially began to assume he was perhaps the prophesied Messiah who would deliver Judah out of the hands of the Romans. But gradually, that perception faded away when they realized that he was just an ordinary man with no enigma of a great king. But he still he was highly revered by the Jews.

Jesus was born of a woman from the tribe of Manasseh. He was conceived by the Holy Spirit but his earthly father was Joseph. When Mary married Joseph she was only sixteen years old and Joseph was forty-nine. Joseph was married previously and had three grownup children. Joseph's first wife died ten years earlier and lived with his mother because his children had married and were living separately. His mother advised Joseph to get married before she died. As a Jew, Joseph searched for a wife among the Jews but was unsuccessful. Even though Jews were prohibited from marrying women from the lost tribes of Israel, many of them did because they couldn't find wives from among the Jews, especially when they were older and advanced in age. Ultimately, Joseph betrothed a beautiful damsel named Mary.

Joseph was a carpenter and Jesus assisted him when he grew up from the age of nine. When Jesus was fourteen years old Joseph died and Jesus became the bread winner of the family. He couldn't run Joseph's shop because he did not have much knowledge of carpentry. Jesus mainly stitched mattresses for the beds Joseph made and occasionally helped in making chairs and beds. Hence, Jesus took jobs at construction sites and became responsible for making doors, windows and fences because he has some knowledge of carpentry and had the tools to make them. He earned barely enough for sustenance and sometimes supplemented his income by making mattresses at home. Throughout his life Jesus spoke to the Lord and was in constant communication with God. The Lord always told him to wait for his appointed time and he complied for he knew that God had a special design for him.

At the age of sixteen, Mary found him a wife to wed but Jesus told her that the Lord rejected her. She continued to look for another woman from amongst her tribe. She found one eventually and Jesus loved her and the Lord endorsed her. But the Lord told Jesus that he will not have children and Jesus accepted his fate. He told his mother and she was grieved but she believed in the Lord and understood that she will never have grandchildren from Jesus. Jesus continued to speak to God and was consistently guided by the Lord. His wife, Amira, also became a devout believer in God and ultimately became a holy woman. After four years she was sad that she did not bear a child because Jesus never disclosed to her the reason but he always disclosed his love for her inspite of her bareness.

At the age of twenty, the Lord appeared to Jesus and instructed him to roam the towns and cities of Judah and heal those that were tormented with diseases, inflicted with disabilities and possessed by demons. He did so for ten years. During the ten years Jesus has a leather sac which he placed in front of him while he healed. After healing hundreds at a time with a simple wave of his right hand many of the healed inhabitants dropped whatever they could afford on Jesus' sac.

And that is how he survived. He had enough money to give the family he stayed with for his food and other necessities and for the caretaker of the donkey he travelled upon. He also saved enough money to send to his mother regularly. He returned to Nazareth more than twenty times during his ten years journey. When Jesus first returned home after six months he realized that Amira was absent. When he inquired from his mother Mary told him that she left to devote her life to God and serve in a monastery. Jesus said that it was the best decision she could ever make in her life. Even though he knew where she was he never ever went to see her and she never ever attempted to contact him. She would ultimately live to an old age and continued to serve the Lord throughout her life and help the old, the infirm and other needy members of the society.

From the age of twenty to thirty Jesus was filled with the Holy Spirit and saw angels surrounding him. At the age of thirty Jesus was baptized by John the Baptist and the spirit of the Lord descended upon him and he became the replica of God. He became God in human flesh. His body, heart, mind and soul became that of God. And from then onward every word he uttered from his mouth were that of God. He no longer healed, drove out demons and forgave sins in the name of God but in his name.

Jesus primarily preached among the lost tribes of Israel. He preached powerfully for hours and told the remnants of the thirteen tribes that they were the authentic Israelites and the chosen people of God. He discarded the Jews as illegitimate children of Israel and compared them to foreigners. The Jews heard what Jesus was preaching and began to send spies to monitor him and what they learnt disgusted them. Jesus was brutal in his attack of the Jews and demeaned them whenever he could. Jesus even placed the Gentiles higher in hierarchy and above the Jews in the eyes of God.

They couldn't understand why he suddenly became and spoke against the Jews. They eventually conspired to destroy him. The remnants of the house were alive for the first time in centuries and ascertained their authenticity as the rightful heirs to Israel. They began to see the Jews as enemies and foreigners occupying their land and challenged them and told them to go back home to Babylon. They easily identified the Jews by their large ears and broken noses and smoother hairs apart from their whiter color of skin. To the Jews, Jesus was the greatest threat in their entire history since their return from Babylon. Jesus was not only a threat to their ultimate authority in Judah but their very existence as Jews. Killing Jesus and stamping out the movement he started was a matter of survival to the Jews. Either Jesus died or the Jews perished they perceived. Ultimately, they succeeded in eliminating him and were crucified and died a very painful death.

When Jesus died the movement he started also died with him, at least that is what the Jews thought. But they began to hear rumors among the remnants of the house of Israel that Jesus was seen alive by numerous people. The Jews began to persecute those that were spreading the "false" rumor. They tried to enlist the help of the Romans but when the Romans inquired they found out that the people merely believed in a dead man to be a god and saw nothing to assume they were a threat to the Roman Empire whatsoever. When the eleven disciples of Jesus Christ saw that Jesus had risen from the dead they were inspired to continue the movement he started. Peter was eventually arrested and was slotted to be killed but was saved miraculously. He escaped to Europe and hid himself from the Jews for a bounty was placed upon his head.

Mary and her son James lived and followed Jesus during his last two years on earth. When Jesus was arrested his mother and other women run to the Roman court and witnessed everything that happened to Jesus from the time he was tried until the time he was buried. Only John did so from all the other disciples of Jesus Christ. When Jesus was being nailed to the cross John, Mary the mother of Jesus, Mary

Magdalene and other women watch from a distance of about thirty meter. They continually wept and some poured dust on their heads to express their sadness. When Jesus was finally raised his feet were about two feet from the ground. Jesus was in the middle and the other two thieves were crucified facing each other and a few feet from the cross of Jesus. All the three could see each other and were at close proximity. Suddenly, Mary Magdalene run from the gathered crowd and headed towards the cross of Jesus. The Roman soldier whipped her and threw her on the ground. She got up again and began to walk towards the cross and the Roman soldier did the same thing and dragged away unto the crowd. She got up for the third time and again advanced towards the cross.

This time the Roman commander stopped the soldier and wanted to see what she was up to. Mary walked barely two feet from the body of Jesus and stood and clasped her hands in a prayer mode and began to console Jesus and pray. She touched the nails on Jesus' feet and had his blood on her fingers. She stood still for hours until she almost collapsed and eventually was forced to sit down. When Mary the mother of Jesus and others saw that the Romans did not retrain Mary Magdalene they took heart and got much closer to the cross. They were all merely three meters from Jesus surrounding the cross for most of the time.

After the resurrection of Jesus Christ his movement began to spread fast. The disciples preached in secrecy and had many followers. One who became an ardent follower of Jesus was Luke. He was with the disciples for sometime and began to see the movement that Jesus began dwindling and gradually fading. They he heard of a man called Paul who was doing a marvelous work among the Gentiles. He left Jerusalem and joined Paul. Paul was a learned Jew and a powerful speaker like no other before him. He was unbelievably fearless man and his faith in Christ was beyond comprehension.

During his only two years of preaching Paul was beaten mercilessly several times and the many scars on his face was a proof of it. His back was full of whipping stripes and had his two front teeth knocked off and he suffered back and shoulder ache throughout his life. In two years of his preaching Paul converted more than a hundred thousand Gentiles to Christianity. Ultimately, he would make the mistake of returning to Jerusalem to consult and advise Jesus' disciples to accept Gentiles as their equals. He will be imprisoned in Jerusalem and ultimately find himself in a Roman jail. He was finally tried and found not guilty by a Roman judge even though the Jews labeled very convincing case against him. When paid was rendered free by the judge the Jews began to shout in discontent and all of them vowed not to enter a synagogue again until Paul was dead. The judge hearing their threats remanded Paul to stay in jail until he was relocated to a safer place because he was a Roman citizen. Many places were suggested to Paul in remote areas of the Roman Empire but Paul rejected them all. He just wanted to be free and preach wherever he wanted and did not care if he was killed in the process. However, Paul stayed for another five years in the Roman jail.

At the beginning Paul stayed in jail and did not have contacts with many people. His followers came to see him regularly. After about two months the judge gave permission for Paul to venture outside the prison yard with two soldiers escorting him for protection. Many people came and he preached outside not far from where he was imprisoned. He did so for almost a year. During that period of time the Roman soldiers began to believe in his teachings. He baptized them by sprinkling a little water on their heads and drawing the sign of the cross on their foreheads. As time went on Paul began to preach further and further from his prison proximity. Two years later his followers came to pick him up from prison and preached in almost every corner of Rome and was converting many. He also began to write numerous letters to the many churches in the Roman Empire. The Roman soldiers stopped escorting him because Paul was always in the company of many followers and they did not fear for his safety any longer. But after four years in prison Paul felt safe to walk out of the prison by himself and meet his preaching partners at the appointed venues.

One day, Paul was walking back to prison and was approached by two men. They accused Paul of disseminating false messages. They argued with him for a little while and suddenly one of them struck Paul badly that he began to bleed. They left Paul on the ground and run away. His followers were very furious when he told them about the two men. After two month Paul recognized the two men who assaulted him among the crowd when he was preaching. He pointed them to his followers just to tell them who they were. However, his followers apprehended the two men and beat them mercilessly. Paul was alarmed by their actions and expressed his displeasure to them. The two men will ultimately report their misfortune to their rabbi who happened to be one of the Jews who attended Paul's trial five years ago. Immediately, he sent messages to Jerusalem alerting them that Paul was preaching in Rome. The Jewish rabbis will hire hit men and when Paul was returning to prison one day they followed him. They had staked him for sometime and knew his routines. One of the men stopped Paul as if to speak to him and while he engaged him the other came out of hiding and struck Paul with a dagger on his back. Paul gave a loud cry and fell to the ground. They beheaded Paul and took his head with them to show to the rabbis. And that is how the life of Paul ended. Paul converted more than ten thousand people in Rome. And through his letter more than two hundred thousand Gentiles became the followers of Christ.

Meanwhile in Jerusalem the church was not advancing and the memory of Jesus was beginning to fade away. The disciples did a lot of preaching but were not converting too many unlike the churches among the Gentiles. Eventually they agreed to write the teachings of Jesus and document their experiences with him. They all agreed to meet every Friday to document every event that prevailed during the time of Jesus Christ. One at a time they mentioned various incidents and when each one of them did he was asked to write the details. All of them recorded separate incidents and no story of Jesus was duplicated. All the ten disciples wrote a gospel very different from each other. It took almost a year to compile all the ten gospels, all unique and different from each other. They then began to read the gospels in the churches and soon many people began to relate to Jesus and many were joining the church. After Paul was imprisoned in Rome, Luke returned to Jerusalem and joined the church and spent ample time reading the scrolls upon which the gospels were recorded.

Six years after the resurrection of Jesus Christ the church was beginning to have a new life. Many churches were established among the lost tribes of Israel and their numbers were increasing steadily. The Jews took note and were determined to wipe out the church once and for all. They enlisted the help of the Jewish militia and knew precisely how the church activities were being conducted and identified key members of the church. One Sunday, the Jews encircled five prominent churches and killed all the men that were gathered to worship, including nine disciples of Jesus Christ. Only John survived and eventually will escape to Greece and live the rest of his life in a monastery. That same night, the Jews went from house-to-house and killed all the men they identified as members of the church. All in all, two thousand Christians were killed by the Jews in just a single day and night. The Jews were very pleased and believed they had wiped out any traces of Christ and would live in peace on their land with their ultimate authority intact. But their irrational action would backfire on them terribly.

There was a great uproar among the lost tribes of Israel over the death of their compatriots and there was a unanimous call for vengeance. The Jews found themselves threatened by the entire children of the thirteen tribes of Israel. There were five thousand Christians when the Jews killed the disciples but within two years after the massacre their numbers increased to thirty thousand. The Jews stood idly as they watched Christianity sweeping and spreading like wildfire among the lost tribes of Israel. They built many churches and dared the Jews attack them and the Jews never did. After thirty years all the remnants of the house of Israel were Christians. The remnants of the house of Israel saw themselves as the authentic children of Israel and considered the Jews as foreigners and heathens because they did not believe in Christ.

After the eleven disciples were killed Mary and James returned to Nazareth. After the crucifixion of Jesus the disciples took good care of Mary and James and provided them with all their necessities. After their death Mary was fearful for her life and kept a very low profile and did not want to be discovered by the Jews. She lived with James, his wife and two children. All her children were dead except James. Ultimately James' wife would die and a year later also James would die at the age of fifty-five. Mary always spoke about Jesus from the time of his childhood until he died. Her ten grandchildren were sick of her stories because Mary never spoke about her other children as if they never existed. They all resented Mary.

When James died, Mary had no one to look after her. None of her grandchildren would take her in or even aid her. They all said, "let the dead Jesus take care of her." As a result, she lived in abject poverty for two years and survived on alms. Her grandchildren never believed in Jesus and followed the Jewish faith because their grandfather was a Jew. Mary's rendition of Jesus' birth they felt was a total fabrication and accused Mary of committing adultery. They were a wicked set of children and ultimately none of them would amount to anything. Only one of them would marry and had a child but the rest became very worthless.

One day, the priest of the church where Mary attended regularly gave a moving appeal and described the pathetic condition under which Mary was living. One of the listeners was a young rich man and was appalled by what he heard. He couldn't believe that the mother of the Son of God was living like a beggar. He took Mary from the church and put her under his roof. He gave her a very spacious room and furnished it with all the amenities she required. He had many servants and slaves and Mary never washed her clothes or prepared food for the rest of her life. Everything was provided for her and she lived like a queen from then and thereafter. The rich merchant showered her with gifts and whenever he raveled he always brought something for her. She pleaded with him not to buy her anymore things but he just wouldn't stop. As a result she gave most of it to poor neighbors. Mary had twenty pieces of clothing and ten pairs of shoes and many shawls to cover her head. Mary never had more than three pieces of clothes all her early life and just a pair of shoes at a time.

The rich man grew in wealth and he truly believed that it was as a result of the blessings of God for sheltering Mary. One day, during a conversation the rich man identified his tribe to Mary. Mary told him that she was also from the tribe of Manasseh. That made him even love her more knowing that she was from his tribe. Mary spent the rest of her years in prayer. She went to church daily and spent the rest of her days visiting neighbors and narrating the story of Jesus which she was never tired of telling. Mary lived a happy and content life. She missed Jesus and her children but her heart was always engrossed in prayer and in the spirit of God that she never despaired. The unique thing about Mary was that she never ever fell sick even for a day throughout her life. She lived to be ninety and never lost her mind nor needed assistance from anybody. Her last year she walked with the support of a stick and dragged her feet when she walked. She still went to the washroom and bathed by herself. At the end she died of old age. Her heart just stopped pumping as she sat on her favorite chair. Her funeral was attended by a very huge crowd that alarmed even the Jews and was befitting of a queen.

When the ten tribes of Israel converted to Christianity they were hated even more by the Jews. The Jews consulted with each other and agreed to drive all the remnants of the house of Israel out of Jerusalem. And soon Jerusalem was exclusively inhabited by Jews only. The remnants of Israel built homes in the open fields away from Jerusalem and established themselves. Trade and commerce was still in the hands of the Jews and the remnants of Israel depended on them entirely. The Jews sold their goods at exorbitant prices to the Israelites and made life difficult for them. Hence, the whole of Judah was divided between the Jews and Israelites and never mixed with each other. The entire city of Jerusalem was inhabited and under the control of the Jews and no Israeli lived in the city. There were two distinct societies that did not interact with each other. However, the Jews had still control over the remnants of the house of Israel because all the Roman militias were Jews. This status quo went on for twenty years.

The Jews were alarmed at what they saw develop among the remnants of Israel. Trade and commerce was flourishing and many buildings rising among the remnants of the house of Israel. Israeli merchants travelled to Egypt, Syria and Lebanon and brought essential good and established their shops in their territories and relived themselves from relying on the Jews. Business thrived among the remnants of Israel and the Jews were jealous and felt that they were gradually losing their grips on the remnants of the house of Israel. Finally, the remnants of the house of Israel would reject the Jewish militia and their elders went to the Roman governor and expressed their displeasures with the Jews. When the news got to the Jews they were very furious at the audacity of the remnants of the house of Israel. The situation got so tense that violence was very imminent between the two divided groups.

In 68 AD the Jews became violent towards the remnants of the house of Israel but the Israelites did not change from their stances. The remnants of the house of Israel rejected the Jewish militia in their entirety and refused to cooperate with them and disobeyed their orders. The Roman governor decided to hold a "reconciliation conference" to pacify the two warring factions of his territory. He formed a committee from among his officers and invited the leaders of the Jews and remnants of the house of Israel to strike a peace deal among them. When the conference began only the Jews spoke loudly and with anger and the remnants of the house of Israel listened silently. The Jews made it clear to the Romans that the remnants of the house of Israel were foreigners and should have no say in the administration of the land that belonged to them. Then the Roman governor asked, "If they are not Jews who are they then?"

"They are Philistines," a Jewish elder replied.

"Who are the Philistines? From which land did they come to Judah?" he asked.

"Our king David defeated Goliath and they became our subjects since then," the Jews replied.

The governor asked several questions to both sides and realized that the Jews were basing their claims on events that happened many centuries ago. They also gathered that the remnants of Israel had never heard of either Goliath or the Philistines. Both sides were claiming to be the authentic owners of the land and each accusing the other of being foreigners. The governor was disgusted by the forceful demands of the Jews to the extent of belittling his powers. The Jews were very adamant in their stances and did not care what the Romans thought and were determined not to relinquish their authorities over all Jewish lands. The governor was impressed by the calmness of the remnants of the house of Israel and their desire to live in peace with the Jews as their neighbors. Ultimately, the Roman governor would make his final decision and side with the demands of the house of Israel. He withdrew all the Jewish militias from amongst the remnants of the house of Israel and recruited new ones from among them. He appointed two thousand members of the house of Israel to help the Romans administer the Israeli territories. For the next two years the remnants of the house of Israel will be totally isolated from the Jews and live as a distinct territory of the Roman Empire with no link to the Jews.

The Jews accused the Romans of giving vast lands of the Jews to the Philistines, the enemies of Israel. They resented the Romans and virtually stopped cooperating with them and disobeyed their orders. The Jewish militia sided with their people and became hostile towards the Romans. Eventually, the Jews refused to pay taxes and the Jewish militias refused to collect taxes from the Jews. The Roman soldiers were beginning to grumble because they had not been paid for ten months because three quarters of the taxes the Romans collected came from the Jews. Eventually, the Jews became bold and began to ask the Romans to get out of their country. They organized and held a huge demonstration with over a hundred thousand Jews participating and shouted slogans demeaning the Romans and demanding immediate departure from their territory. After that the Jews wrote a letter signed by their elders giving the Romans an ultimatum of three months to evacuate all Jewish lands. The four thousand strong Roman army knew that their days in Judah were numbered. They couldn't possibly fight a hundred thousand Jews. Hence, the Roman governor wrote an immediate report to Rome and was waiting for a response.

After seven weeks the governor was happy with what he saw. The Roman army was reinforced and swelled to twenty thousand soldiers. The governor read the letter from the most senior representative of the Roman Emperor based in Macedonia with a simple instruction that read, "exterminate the Jews." Shortly thereafter the Roman governor of Judah was handed over a letter signed by the elders of the Jews giving him just two days to depart Judah or face the wrath of the Jewish people. The next day, the Roman soldiers marched the whole night and surrounded Jerusalem in its entirety. At the first sight of the sun they began their slaughter of the Jews and did so until dusk. At the end of the day, out of the three hundred and twenty-five thousand Jews that lived in Jerusalem and its environs only twenty-six thousand remained alive. None of the adult men and women that the Romans encountered were spared but only young children survived. Those that survived run unto the mountains and hid themselves from imminent death. For the next two days all the remnants of the house of Israel were ordered by the Romans to bury the dead Jews. Most were buried in shallow graves and some thrown into a nearby ravine. The remnants of the house of Israel did not express any sympathy for the Jews and felt they deserved what they got.

The surviving Jews would finally return to Jerusalem. All other Jews from every corner of Judah would converge and live in Jerusalem and inherit the properties of their fallen countrymen and women. There were only thirty-five thousand Jews left and they were all in Jerusalem. They gradually consolidated their powers and enforced their authorities while vowing unflinching loyalty to the Roman emperor. However, their resentment of the remnants of the house of Israel was unprecedented. They hated them because they did not participate and support the Jews that stood against the Romans. They considered them more as archenemies. They made futile efforts to dominate the members of the house of Israel again. The remnants of the house of Israel were surprised at the audacity of the Jews and were repulsed by their conducts and thought of exterminating them just as the Romans did.

One night the Romans were overwhelmed by weeping and traumatized Jews who urged them to save their fellow Jews. The Romans had no idea what was happening in the city of Jerusalem and were unwilling to find out until sunrise. By morning almost three thousand mostly young Jewish men and women had sought sanctuary among the Romans. When they marched into the city in the morning what they saw shocked them. Bodies of men, women and children were strewn all over the city. When they saw the Roman soldiers some Jews began to emerge from their hidings. About three thousand more Jews joined those that were in the Roman camps. All in all, a total of five thousand Jews survived and thirty thousand Jews were slaughtered by the remnants of the house of Israel. The Romans marched to the territories of the remnants of the house of Israel and demanded that they handed over the culprits that were responsible for the genocide. No one came forward and all expressed ignorance. The militias were able to arrest only twelve people that had blood on their clothes and after interrogation only one of the arrested pointed fingers on three more participants. All confessed that it was too dark and did not know who actually was next to them. The Romans ordered a sentence and the fifteen were hanged immediately.

The Romans threatened to exterminate the remnants of the house of Israel if they ever dared touch the Jews. They urged the Jews to return to their homes and promised to roam the streets of Jerusalem by night to protect them but the Jews rejected their offers. All the Jews stayed in the Roman camps, scattered all over the fields, for the next two months. The Romans forced the remnants of the house of Israel to provide food for the Jews that were in their camps. Then the Governor wrote a detailed letter to Rome and waited for instructions regarding the position of the Jews. He got a response from Macedonia and was instructed to disperse the Jews in remote areas of the Roman Empire "very far away from Rome." The governor took out his map and picked Portugal. The governor did not have the money to transport five thousand Jews to Portugal. When he inquired he discovered that most of the Jews had money stashed in their homes. Luckily and after the massacre none of the remnants of the house of Israel had returned to Jerusalem and all the Jewish homes were intact. Hence, the Roman soldiers escorted the Jews and retrieved their monies from their homes. They also went to all the vacant homes and searched for money. The surviving priests went to the synagogues and preserved some Torahs and other ornaments they could carry to their new settlement. In essence, almost all the Jews paid their own fares to Portugal. It took almost five months for all the Jews to be evacuated to Portugal. But some of the ships got lost and landed in Spain and offloaded their human cargo there. A Roman soldier accompanied every ship that left with the Jews with a letter indicating that it was an order from Rome that the Jews should be resettled in the Roman territory. By 75 AD there was not even a single Jews in the entire territory of Judah.

The remnants of the house of Israel, all of whom were Christians, established themselves in Jerusalem and other cities and became the ultimate bosses and flourished as a people. The Romans called them Palestinians and they did not object even if they had no clue of the origin of the name. However, the Palestinian knew that they were Israelites and worshipped God in the name of his Son, Jesus Christ. And so life continued for the next fifty years. One day, a proclamation came from Rome declaring Christianity as an illegal religion in the Roman Empire. The governor came out publicly and threatened execution for every Palestinian that was caught praying and mentioning the name of Jesus. The militias quickly went to work. All the churches were sealed and the Palestinians complied with the orders of the governor and never preached or propagated anything that was Christian. Within just twenty years the name of Christ would fade away but the Palestinian continued to believe only in God even though most of them identified themselves as Christians.

Twenty years after the ban on Christianity was enforced in Palestinians the Roman governor realized that the population of the Palestinians was dwindling. When he inquired he discovered that about a hundred thousand Palestinian had migrated to Egypt to worship God in the name of Christ. He was very furious and decided to take drastic measures. He ultimately arrived at a decision that he would regret terribly. One day, the Palestinians saw all the Roman soldiers advancing southwards and didn't know where they were heading to. They returned ten days later and what they saw shocked them all. As the Romans marched towards their camp they discovered that they were fewer in number and virtually all of them had sustained wounds all over their bodies and some were barely clinging to life. Apparently the Roman governor, in his fury, made an irrational decision to punish the fleeing Palestinians that had sought sanctuary in Egypt and betraying Rome. A century earlier the Romans had conquered Egypt but after some years the Egyptians rose against them and wiped them out and Rome never attempted to dominate them again. Hence, when the Palestinian governor marched to Egypt he was faced with a rude awakening. The Egyptians rose collectively and fought the Roman soldiers and were overwhelmed by the huge multitude that attacked them. They fought valiantly but couldn't overcome the enormous number of Egyptians that fought against them. Finally, they were forced to withdraw but not before majority of them were killed. Out of the five thousand Romans that marched to Egypt only a thousand and two hundred were able to escape with their lives.

When the Palestinians realized what had happened they were encouraged by the actions of the Egyptians. Two days later they all stormed the Roman camps and finished all the weakened Romans. They broke into the arms storage and armed themselves ready for any Roman retaliation. But the

Romans never came again. Rome was experiencing revolts in many parts of its empire and Palestine was not a priority and as a result never sent soldiers to reoccupy it.

Only five thousand of the Palestinians who migrated to Egypt will finally return. These Egyptian-Palestinians will continue in the right way of Christianity and become faithful followers of Christ. The rest of the Palestinians were just believers in God. A century later the faithful Christians will be ignorant of the messages of Christ and only know that a man died and rose from the dead. The other Palestinians ever went further astray. They learnt that some of their elders saw the risen Christ and described him as an ugly and scary creature that had horns, hoofs and other despicable features. They assumed that Christ became the devil and they believed it for many centuries. The minority faithful Christians disputed their claims but were threatened and sometimes killed. The Palestinians had no Torah, copies of the Bible and didn't have scribes that were learned in Aramaic or Hebrew. The Aramaic alphabets disappeared because no one wrote in them.

The Palestinians will grow in number and have absolute control of the territory of Judah. They became so powerful that they did not tolerate any challenges from their neighbors. They only believed in themselves and did not create any form of alliance with their neighbors. No trading with their neighbors. They were especially hostile to Gentiles. They ensured that no Gentile entered their territory. Christians from Europe who desired to go on pilgrimage to Jerusalem were deleted. They normally disboweled them and they died a painful death. They were always suspicious that the Gentiles would occupy their land again and thus the extreme hostility.

After five hundred years the Palestinians were divided into a hundred clans all over Judah. Each clan had its own dialect by which they were identified. They all spoke different dialects derived from Aramaic and no one spoke the original language. Gradually, the clans began to be hostile to each other. They attacked each other mainly to confiscate women from neighboring clans. Over the centuries they extended their invasions to the neighboring nations. Palestinians invaded Lebanon and Syria and seized women and properties and became a torment to their neighbors. As a result, the areas of Syria and Lebanon neighboring Palestine were rendered desolate and were not inhabited for fear of the Palestinians. The Palestinians became evil, wicked and despicable people.

At the end of the 1100 AD the Palestinians attempted to invade Lebanese territories but were repulsed and driven back with so many of their men lost. The Syrians were even worse. Whenever they attempted to invade the neighboring towns of Syria they were completely annihilated. They didn't know how the Lebanese and Syrians became so brave and acquired superior weapons. They one day, in 1192 AD, a large numbers of Syrians entered Palestinian territory and began to occupy their lands. The Palestinians were divided into a hundred clans and had no unified army. The advancing Arabs conquered a few clans and wiped out anyone that was on their way. Seeing the hopelessness of the situation the Palestinians surrendered to the advancing Arabs. Soon, more than twenty thousand Syrians Arabs preaching a new religion called Islam dominated Palestine. The Arabs discovered that the Palestinians believed in God but that they described the prophet Jesus as the devil a belief that angered them. All the Palestinians, with the exception of about five thousand devout Christians, were converted to Islam with no problem. The Arab went even further. They kept all the young Palestinian women as concubines. Some Arabs had up to ten concubines and there were no women left for the Palestinians to marry. Only the older Palestinians men had wives. A century later Palestine will be inhabited primarily by Arab bastards. All traces of the original Israeli inhabitants will disappear. These Arab bastards will become the forefathers of the present day

Palestinians. Only the few Christians that survived for many centuries will be full-blooded Israelites.

The Arabs in Palestine will assert their authority and become the new owners of the land and portray the zeal and tenacity of their vicious ancestors. They did not have enemies and lived in peace with their neighbors. Then the crusaders came. Initially, they welcomed them and about twenty thousand of them made it to Palestine in three years. The crusaders stayed for ten years before the Palestinians realized that they were determined to live on their land permanently. The crusaders began to challenge the authority of the Palestinians and they were saddened. One day, they gathered their forces and fought them. In the war, thirty thousand Palestinians and ten thousand crusaders were killed. The rest of the crusaders surrendered and were given clemency and were not killed by the Palestinians and ultimately melted into the Palestinian population. That is why some Palestinians still have some European features.

The evacuated Jews made it to Europe with no significant incident. All the ships they boarded made it safely to their destination. Four thousand and five hundred settled in Portugal but five hundred of them landed in Spain because of the weather condition. The Roman governors in Portugal and Spain welcomed them and did everything possible to resettle them because they were instructed to do so by Rome. Jews began to flourish in both countries and multiplied in great numbers. By 1600 AD their numbers had grown substantially. There were five hundred thousand Jews in Portugal and three hundred thousand in Spain. Their numbers would have become even more if many of them had not converted into Catholicism in both countries.

In Portugal, the Jews were prosperous and indulged more in building homes and acquiring land properties. They were so dominant that virtually all the good buildings in all the major Portuguese towns and cities belonged to Jews. The local inhabitants were reduced to mere renters. The Jews conspired together and raised their rents to the level they desired and the locals struggled hard to keep up with the payment. Those that delayed the payments of their rents were thrown out of their homes with no mercy. And the people began to grumble a lot. Most of them paid rents but had very little left for food and other necessities. Life was very hard for the local Portuguese inhabitants living in the major cities and towns. Only those living in the rural areas had control over their lands and homes.

In Spain, the Jews dominated in a different sector of the economy. Trade and commerce ultimately fell into the hands of the Jews. Jews became very rich and the wealth of the Spanish nation was virtually in the hands of the Jews. Even the king of Spain was under the mercy of the Jews and he feared to offend them. The Jews started by undercutting prices to destroy their competitors. Whenever a Jew started a business he sold his wares at a much lower price, even at a loss. As a result, competing business lost clients and eventually went bankrupt. When all the competitors were closed the Jew doubled or tripled his prices and the locals had no options and alternatives but to buy from him. That way they were able to dominate all business transactions and owned most of the outlets. Again, as in Portugal, the Jews primarily dictated the prices of goods sold in Spain.

Portuguese and Spanish people expressed their dissatisfactions to their leaders but the Jews were seen as valuable assets to the nations. The Catholic churches in both Spain and Portugal were earning very little revenue from the citizens and the Jews never contributed a cent to the churches. Churches began to abandon their parishes for lack of funds. The Jews had a stranglehold on their economies that the local people had no extra money to donate to the churches. Ultimately, the message got to the pope in the Vatican and was told of the tragic situation of the churches in Spain and

Portugal. He understood that the problems were the Jews. Hence, and after verifying the prevailing facts in Spain and Portugal, he signed an edict ordering the leaders of the two countries to deport the Jews from their land.

He Jews in Portugal had full control of every facet of the country. They virtually owned Portugal. There were over five hundred and fifty thousand of them and with their hired servants they were almost two million strong. They did whatever their hearts desired in Portugal. Some of the radical Jews even went and forced their way to the church pulpits and preached to the Christians. They fearlessly described Jesus as a liar, a demented and sick human being and a great deceiver. They used whatever derogatory terms they perceived in their hearts in describing Jesus. The rulers and church leaders were afraid of the Jews and no one contested against them. In Spain the situation was different for many centuries and anyone who spoke against Christ, the church or the pope was hanged. But the years before their deportation the Jews became increasingly daring and defied church and government laws. They spoke openly against the church and also insulted Jesus using various evil terms. The Jews in Spain and Portugal were fully convinced that they could ultimately destroy the church and permanently discredit Christ. Many of the Jewish generations never forgot what the Romans, with the help of Christians, did to them in Jerusalem and how they were massacred and driven out of their lands. The pope ordered their deportation from Spain and Portugal primarily because of their open hatred and contempt for Christ.

In Spain, the pope's order began to take effect two days after the king of Spain made the proclamation known to the people. Jews were immediately rounded up and placed in various concentration camps to await their deportation. No Jews were allowed to take anything other than the clothes on their backs. Men, women and children were lined up and marched to the eastern border of Spain and told never to ever return to Spain. The Spaniards were highly organized and had a large army and the soldiers took the full responsibility of deporting the Jews. The soldiers were rude and cruel and beat many of the Jewish men on their way to the camps. The local inhabitants also cooperated fully and sniffed out many Jews who went into hiding. Within just ten days there wasn't a single Jew in Spain. Many Jews wept in churches and pleaded with the church representative to appeal to the king to have mercy upon them. Many even swore to convert to Catholicism but none of the priest had the power to disobey the pope or the king and could do nothing for them.

The Spaniard Jews marched eastwards being led by some merchants who knew their ways across Europe. Many of them marched for a couple of days and their legs were tired and had sore feet. Gradually, the most elderly just sat by the streets and could not walk any further and they will die eventually of starvation. Many women just dropped their infants by the side of the roads and marched forward because they could not carry them any longer. It was a matter of survival. Many Europeans were alarmed by the sudden invasion of Jews. Seeing their tragic conditions many of the towns and villages stood by the roads and gave bread and whatever they could afford to the advancing Jews. But then the flooding of Jews became endless and the local European inhabitants exhausted whatever they could spare. Whenever they learnt that the Jews were ordered deported from Spain by the order of the pope they didn't welcome them in their midst. They were forced to continue their journey. Out of the three hundred Spaniard Jews deported about thirty thousand of them died mostly the elderly and young children.

In Portugal the Jews heard the orders of the pope and the instructions passed by the Portuguese leaders and they dared them touch any Jew. The Portuguese did not have a large army and the government was very weak and could not carry the orders of the pope. It took twenty days before the first Jew was deported. Initially, the Jews discarded the orders of the pope and threatened the rulers with reprisals. But the church, especially, began to organize volunteers and quickly established local militias that were given the responsibility to apprehend and gather the Jews to specific areas. The Jews initially ignored their threats but when they learnt of what happened to their fellow Jews in Spain they began to take the matter seriously. And when they saw the entire Portuguese local population organized into militias they saw the writings clearly written on the wall. Hence, instead of attempting to stay in Portugal they began to prepare for migration. The Jews owned over fifty thousand horses out of the two hundred thousand horses that existed in Portugal and more than two thousand carts and carriages. Hence, more Jews began to buy more horses and carts. The locals suddenly tripled the prices of horses and the Jews had no other options but to pay for them. They added ten thousand more horses and three hundred carts to their assets.

The Jews in Portugal were not brutalized and had ample time to prepare. Many of the Jews went to the various churches and expressed their desires to convert to Christianity. Unlike Spain, the Portuguese heeded to the pleas of the Jews. Some Jews who did not own properties and lived in harmony with the local people were baptized and given a certificate of baptism. All they had to do was bring three friends or neighbors who vouched for them. The witnesses were sworn by the priest to ensure they were telling the truth. Many rich Jews paid witnesses to lie on their behalf in order to obtain a certificate of baptism. They gave the churches huge sums of money but they were betrayed by the witnesses when they realized that they had to swear to an oath. But ultimately almost forty thousand Jews were baptized and became Christians and were not deported.

When the Jews were ultimately driven out of Portugal they had some provisions and collected some essential amenities from their homes and the local militias did not confiscate it from them, unlike in Spain. They had three carts loaded with over one hundred Torahs. However, they were strip searched to ensure that they did not possess any gold, silver or currency. Hence, the Jews marched eastwards and across Spain and into neighboring European nations. They had tens of thousands of horses and carts under their disposal and they packed their wives, children and the elderly on them. Three to four children rode on each horse and so also did the adults. They took turns in riding the horses and constantly exchanged places in the carts. The rabbis had designated horses and never walked for a day. No Jew died on the way except the few that had chronic illnesses. Unlike the Spanish Jews the Europeans they encountered were not very friendly to them and gave them very little food, if any. They asked them to leave their lands because they had enough Jews in their midst already and were afraid that they will overwhelm them eventually. Hence, the Portuguese Jews marched further and further to the east until many of them finally reached Germany. On the way, they nearly starved to death and many had sold their horses and carts and hence, were forced to walk all the way to Germany. Many of them survived by chewing leaves from plants and even ate grass to sustain their lives. Many got sick but none died.

The Jews from Portugal were smarter and learnt a lesson from the experience of the Spanish Jews. As they advanced eastwards they never identified themselves as Jews and never said they were deported from Spain or Portugal and never complained about the pope and his "cruelty" to uproot them from their lands. None of them identified their place of origin as Portugal and Spain but mentioned other European nations. When they got to Germany every Jew was named as Adam, Abraham, Isaac, Moses, Jonah or other biblical names. No Jew mentioned his or her Portuguese or Spanish name. After many years all Jews will exclusively adopt German names and very few would retain their original names. But virtually all of them discarded their Spanish and Portuguese names and didn't want to be identified with those two "evil" nations again and none of the Germans knew where they migrated from. The Jews would later change their tones and would begin to say that they migrated from Jerusalem and the Germans believed them.

The Spanish Jews scattered all over Europe and would end up as beggars, surviving on alms. But the Portuguese Jews advanced much further east until they reached Germany. They told the Germans that they had to reluctantly depart from their land because of drought hat afflicted their land. The Germans assumed that they were in their country just to weather away the bad time and would ultimately return to their country. Hence, they began to hire them as farmhands and soon many German farmers were scrambling for them. The Jews were hard working and labored with their wives and children for the farmers. The farmers paid them very little and were mainly responsible for feeding them. They were no better than slaves. Many of the German farmers expanded the land they tilled because of the extra hands they had and became very rich. And that continued for twenty years. There were over four hundred thousand Jews in Germany at the time.

After twenty years in Germany they noticed other Jews from Western Europe advancing further eastwards and never settled in Germany. When they inquired the German Jews discovered that the eastern European countries were very hospitable to Jews. Jews immediately acquired lands and integrated into the society quickly and lived as equals with the local inhabitants. That was especially true with Russia. Hence, the German farmers began to lose their farmhands and didn't know why they were disappearing from Germany. When they realized what the reason was all the farmers began to allot small pieces of land to the Jews and pay them higher wages because they simply couldn't sustain their farms without the help of their Jewish workers. The flood to Eastern Europe subsided dramatically and the Jews in Germany were much happier.

In just fifty years since their entry to Germany the Jews became an integral part of the society. Most migrated to the towns and cities and established themselves as full fledged Germans. They became a well-organized group of people and had many synagogues and were a proud set of people. They had learnt a great lesson in Spain and Portugal and never spoke against the church, the pope or Christianity. They did not underestimate the influence of the Catholic Church and the enormous powers of the pope who changed their destiny with a single stroke of a pen. As a result, the Jews gathered money from their members and donated it to their local churches and the Germans loved them. The priests began to preach that the Jews were the chosen people of God and the Jews loved them.

For more than three hundred and fifty years the Jews will live in Germany in perfect harmony. They had the rights and liberties just like other white Germans. They were never discriminated against. The Jews were, however, known for their unity. All Jews spoke in one voice on issues affecting them especially. Hence, they became a force to reckon with in Germany. But unlike in Spain and Portugal the Jews never controlled and dominated any sector of the German economy. They were just regular Germans. Some were rich and so were many other white Germans. They were farmers, civil servants, regular workers, homeless, criminals etc. just like other Germans. There was nothing special about the Jews in Germany. The only thing different was that the Jews had above their fair share of newspapers. Even then, they were not dominant by any means and there were hundreds of other newspapers owned by white Germans. Intermittently, and over the history of the Jews in Germany, a white candidate spoke ill about the Jews and the Jews ensured he lost the election. They travelled in large numbers to the district of such candidates and registered themselves as voters and defeated such politicians.

Even then the Germans did not see such acts as evil. Instead, they accused the defeated candidate and blamed him for opening his "big mouth" and speaking evil against the Jews. After all, he was defeated in a fair election and they perceived that the Jews had every right to defend themselves. That way, the Jews tamed the white Germans and no candidate isolated or spoke against the Jews throughout their history in Germany. Those who did were almost run out of town and totally ruined thanks to the many Jewish newspapers.

In 1918 the population of Jews in Germany had risen to five million. The relation of Jews with white Germans had not changed an iota. Jews continued to live in Germany in perfect harmony. However, the situation of Germany was tragic. Germany lost the First World War and was in total shambles and its economy in total ruin. Hundreds and hundreds of thousand of Germans lost their lives in the war and almost all German families were bereaved. The government had no money and jobs were scarce and poverty and hunger was written in the faces of most Germans. Things never improved for a long time.

In 1925 a Jewish rabbi and a travelling white German businessman met and had their regular friendly talks in a small hamlet of Germany. After discussing many issues and the problems Germany was facing the businessman asked the rabbi saying, "Rabbi, I am impressed by the zeal of the Jewish people. While white Germans are finding it difficult to overcome the effects of the war the Jews have recovered fast and seem to be doing well."

"The devil misled the German people. They fought a stupid and senseless war and nearly destroyed the country. Jews had no part in this evilness. No Jew participated in the tragic war. That is why we are faring well," the rabbi replied.

"Rabbi, I don't understand. Are you saying that no Jew died in the war?" the businessman asked inquisitively.

"Absolutely!" Not a single Jew. We all agreed not to aid the devil and become his instrument," the rabbi replied.

"I am very shocked rabbi. How could you betray Germany in a time of her utmost need? All Germans should be prepared to shed their blood to defend their country. The Jews are Germans and they should have done the same thing," said the businessman.

"We are Jews and we have our own principles. Our religion forbids us to shed the blood of those who did not mean harm to us. We did not feel threatened by those other Germans perceived to be their enemies," replied the rabbi.

"But rabbi, you are a German first and a Jews next. Your priority should be the interest of Germany first. How could you do that to Germany? It is your homeland. How could you desert Germany in her utmost needs? You are sacrificing your existence and destiny. I am bloody sick," commented the businessman.

"We are Jews and Germans just like other whites are Bavarians and Germans," responded the rabbi to the white businessman's comment.

"Jesus Christ! What the hell am I hearing!" freaked the business and walked away from the rabbi cursing and spitting venom.

Three weeks later, the rabbi was sick when he heard what the people in the hamlet were murmuring amongst themselves. The popular rabbi gradually became an outcast. The white Germans in his community totally ignored him and hardly any of them spoke to him again. Not only him but also other Jews were isolated from their community and hated by white Germans. After ten months the rumors had spread to nearby towns and finally made it to the front page of a local newspaper with a bold writing titled, "Germany betrayed." The article in the newspaper sowed the first seed of hatred towards Jews. And in no time it became national news and inflammatory headlines were constantly printed designating

Jews as "traitors." And the lives of Jews in Germany will never be the same again, ever. Larger newspapers dug into the archives of German army recruits and interviewed prominent German generals and confirmed that indeed, no Jew joined the German army and fought in the war. A total of only twenty Jews were killed during the entire four years war and they were all administrative workers.

In four years time the fact that Jews did not join the army became irrelevant. Some Germans even regretted not doing what the Jews did and could have saved many members of their families. However, the movement that started with a minor discussion between a Jewish rabbi and his white businessman friend will take a much bigger dimension. Germans began to see their country as a nation within a nation. They all agreed that there was separate Jewish nation within their German nation. Jews were ostracized everywhere in Germany. The church was the only friend of the Jews. Priests reminded their members that the Jews were a chosen people of God and that unwarranted hatred towards a fellow human being was a sin. But the people continued to vilify the Jews.

Germans demeaned Jews in all their talks and formulated many bogus stories and attributed them to the Jews. If there was a coffee shop owned by a Jew in a town they exaggerated their concerns beyond proportion and described it as if their town was occupied by a foreign force. If women broke glasses in their kitchens they blamed the Jews for it and felt that the glass was broken because they saw a Jew that morning. But still, Jews were not beaten or killed anywhere in Germany. Germans still went to coffee shops owned by Jews and bought wares from Jewish shops and nothing changed. The only thing was that Germans spent their past time denigrating Jews and it became a tradition and it was just talk and no action. But the days of accepting Jews as full-fledged Germans was over. Every German saw all Jews as foreigners who took German jobs and occupied German lands and the poor performance of the German economy did not help also.

The Jewish newspapers went into full gear. They printed articles upon articles to revive the fraternity that existed between Jews and white Germans. They mentioned many prominent Jews that played important roles in the development of the German nation. Quoted many important German personalities who solidified the brotherhood of Jews and white Germans. Mentioned the names of small towns that were pioneered by first Jewish settlers and how they grew up to be major towns. Printed comments from top church leaders and their desires to see a united Germany and the roles of the Jews in the growth of the Catholic Church. The numerous articles helped but could not stamp out the prevailing sentiments of the German people. Overwhelming majority of white Germans, however, couldn't just see Jews as Germans and designated them as foreigners, once and for all.

In the city of Koln lived a very elegant, well groomed and beautiful woman. At a very young age her father sexually molested her and when he finally died at the age of ten she was very glad and felt relieved and wished he would go to hell and burn to eternity. Her mother was always ill from as far as she could remember and was aware of what her father was doing to her and did nothing to prevent it. Ultimately, at the age of fifteen, she would abandon her ill mother and become a prostitute. She regularly visited five or six bars and provided sexual favors for money. She always ensured that she did not have virginal sexual intercourse with her clients. But after being in the trade for three years she became a patron to a rich German businessman. He regularly took her to his guesthouse and had intimate relationship with her. She regularly accompanied him on business trips and was his regular customer and he provided her with everything she desired and was she very happy.

After sometime she realized that she was pregnant. She told the rich businessman and discarded her and never saw her. She was sad and hence started patronizing the bars again. Six months later her belly was large and couldn't attract clients anymore and was forced to live in her apartment all by herself. Gradually she exhausted all her savings and didn't know what to do. She sought help from her large family but they all hated her. Her only eldest sister, ten years older than her, had no sympathy for her. Her sister hated her because she felt that the woman spoke evil about their father when she revealed to her that she was being molested by her father for the elder sister was never abused by her father. Finally she went to the rich businessman and when he saw her plight he was sympathetic and he began to give her enough money until she gave birth to a very lovely son. She called the name of her son Adolf. The rich man knew that she gave birth to a son but warned her not to ever see him or her son, ever, for he was married a man with children.

At the beginning the woman fed her child properly and went to the bars when she laid him to sleep. She returned after two or three hours and always found him fast asleep. She normally serviced three or four men in a day and what she earned was sufficient to pay the rent and have a decent living along with her son. As winter approached she began to stay for much longer hours in the bars because less people were available that required her services. At the age of six months she always found her son crying, whenever she returned from her work, and wondered for how long he was doing so. She felt very sad and attempted to find help from her family members and none would take in the child. She pleaded with her elder sister and the sister told her that she had two children of her own and had no room for a third one. After going through all members of her family only one of her father's sisters was willing to help. Hence, the child began to live with his mother's aunt and he did so until he was five years old.

The mother visited him three times a week and gave her aunt enough money to take care of him for she was poor. Whenever the mother visited him she showered him with kisses and hugs and was very proud of her son. As he grew older the mother began to take him out and was with her for longer hours and she constantly bought him ice creams and chocolates and whatever the child desired. The child loved his mother very much and was always looking forward to seeing her again. However, his relationship with his great aunt was different. For as far as he could remember in his short life his great aunt slapped him, kicked him like a football, punched him and abused him excessively several times a day and the mother did not know about it and the child never revealed it to her. Until the age of three the child cried his heart off whenever he was beaten and this happened for several times in a day. From the age of three he stopped crying because his great aunt did not stop hitting him until he was quiet. Whenever his great aunt abused him he took it silently and absorbing the pain because he knew that crying entailed more whipping. But still he cried sometimes because he couldn't contain his pain. Hence, the child learnt how to hate from his early age and at the age of five he feared and hated his great aunt with passion. At that age he learnt survival techniques comparable to an adult.

One day, the young child asked his aunt, "grandma, I am a good boy." Why do you hit me always when I didn't do any thing wrong?" The great aunt was disgusted by his audacity and replied, "you little bastard. You are a disgrace to our family. You don't have a father and that is why I detest you," and then gave him a dirty slap. His mother continued to visit her son and was happy that he was growing up fast. She always thought that she would one day take him to his father and show him what a beautiful child he had missed.

One day, when the child was five years old, the mother was alarmed by a gush on the side of his head. When she inquired from her aunt she told her that it was just an accident and that he hurt himself when he fell while playing. The mother didn't say anything and she took her soon to their regular outings to the town. While they were sitting in an ice cream shop the mother advised her child to be careful when playing to avoid hurting himself and pointed the open wound on his head. The child replied, "mommy, I didn't fall." Grandma hit me with a stick on my head." The mother was shocked. She asked him if grandma had ever hit him and what she heard from him infuriated her. She was so disgusted. She vented her extreme anger on her aunt and took her son to her apartment that same day.

The mother left the child in the apartment from ten o'clock in the morning until three in the afternoon and he never said anything. In the night he buried himself under his mother and was happy with the weight and warmth of his mother. But whenever he was alone in the apartment he was always fearful and always thought that his grandma would come and whip, punch and kick him. One day the mother was passing by a bookstore and had the urge to buy children's books to teach her son. She bought one book of the alphabets and one of numbers. That evening she sat her son by her side and went through the alphabets asking him to repeat after her. She was not sure if he was absorbing anything. She went through the alphabets over and over for about twenty times. Then she asked her son if he could remember any o them and what she heard shocked her terribly. Adolph was able to name all the alphabets. She asked him at random and he knew all the alphabets. She looked at her son intensely and with amazement and thought he was not a regular human being.

The next day she did the same thing with the numbers and he memorized them in no time. For the next three months she taught him regularly and he was able to read and write and even do simple arithmetic. She bought him more and more books, each time more advanced than the previous ones and he read them with little problem. At the age of five and half he was reading fluently and counting in hundreds. The mother gradually felt inadequate and knew that she would not be able to provide her son with the education he deserved. She knew that he was an exceptionally intelligent kid and a genius and he deserved a better life. Finally, she decided to get him adopted knowing that only rich families adopted children and wished her son to live in a better home and environment. She contacted the nuns in her parish and they agreed to help her son. And so, Adolph was adopted by a family at the age of five and half years. He lived with his birth mother for only six months.

Adolph was adopted by Alfonso Hitler and his young wife and lived in Frankfurt. Alfonso Hitler was a surgeon and worked in a hospital but his wife was a homemaker. Adolph immediately started going to school and was an outstanding scholar and was always on top of his class. His adopted father was very proud of him. Adolph didn't know the reason why he had to depart from the mother he loved and live with his new family but he was not bothered that much. He lived in comfort and his adopted parents were loving and kind to him and never abused him. He was an obedient child and never disappointed his parents for most parts of his early years.

At the age of thirteen he had an incident at school. He fought with a fellow student and he hurt him badly. The matter was reported to his adopted parents and his mother whipped him on his butt with a belt. He didn't cry and understood why he was punished and didn't take it seriously. That night he became so enraged at what his adopted mother did to him and began to hear clearly a voice saying, "evil wicked! Evil wicked! Evil wicked..." From then onward he detested his adopted mother. In the following days he began to see something tragic. He saw a child being kicked, punched, whipped and tormented and the child wouldn't cry. He felt so sad for the child and began to sob profusely. The same child appeared to him for sometime and each time he was very sad. He did everything to remove the image of the child from his mind and finally he disappeared.

Nothing of worth of note happened until Adolph was sixteen. At that age Hitler became a bit rebellious. He became self assertive and his views were not always in agreement with his adopted father and mother. There was total chaos in the house because they never agreed on virtually anything. His father was normally a quiet person but his mother was very talkative. Her anger burst and she spoke at the top of her voice. Even still they never resented him and were proud to have him as their son. His father especially realized that he was passing through puberty stage and never discarded him.

At the age of seventeen he noticed his father sitting down feeling very sad. This continued for more than a week. He finally asked his father why he had such a feeling and the reason why he doesn't talk to them much anymore. His father told him that as a surgeon he saw many patients die at the operating table. However, he just couldn't forget his last patients who died while he was operating on her. He still saw her eyes and pleas begging him to save her life but he failed her and felt so guilty. Adolph understood and said nothing but he knew that his adopted father was depressed. One day, when he returned from school he found two police officers inside his home. Neither his adopted father nor mother was present. After he listened to the officers he discovered that his adopted father had killed his adopted mother. He felt as if the whole world had turned upside down. But for reasons he couldn't understand he felt so relieved, a couple of hours later, and as if he was just released from prison.

His adopted mother's funeral was held two days later and he went to view her body before she was buried. When he saw her body he never shed a single tear and was glad she was dead in his heart. He lived in his home alone for sometime and went to school regularly. He ate food with a rich neighbor who was a lifelong and very intimate friend of his adopted father. Three months later the home he lived in was sold because the defense fees in his adopted father's trial was mounting and his father had exhausted all his savings. The rich neighbor invited Adolph to live in his home for as long as he wanted. He was just glad to be of service to his good friend who was languishing in prison. Hitler initially lived in the main house and was not very comfortable. Then he saw that a room was vacant in the servant's quarters and Hitler asked the rich man if could be allowed to stay there instead. He received his consent and he moved and lived there for the next ten years.

For the next two months Adolph wiped out his memory and forgot that he ever lived with the Hitlers. He felt used and abused by his adopted family. He understood that his adopted family just had him for prestige and to boast among their family and friends and saw himself as a mere trophy that was paraded as their winning gift. The Hitlers were not the kissing and hugging types and never for a single day equated them with his birth mother. All his stay with them was just pretence and never accepted them as his parents. Now that his adopted mother was dead and his father in prison for the next ten years he felt free and liberated and began to have great yearning for his birth mother. His adopted father was craving to see his son in prison but Adolph never attended any of his trials and never visited him in prison. Ultimately, his adopted father would be full of remorse for what he had done and spent days weeping. He would develop one sickness after another and would die two years later and Adolph never ever knew what happened to his adopted father and he never cared.

Two months after the death of his adopted mother Hitler was gripped with the desire to see his birth mother. Since he left her he had her memory stuck in his brain. He remembered most of the details during his short period of stay with his birth mother and she was the ultimate love of his life. When he was living with his adopted parents his father had opened a bank account for him beginning from the age of thirteen and deposited a steady amount of money monthly. When his adopted father went to prison he went to the bank to inquire and found out that his money had accumulated to an equivalent of five thousand dollars. When the urge grew stronger and stronger he travelled in search of his mother. He still remembered the street where they lived in because his mother took him around and asked him to read different signs on the streets. He took transportation and finally found the street. He did not remember the number of her address but could recognize the two story building if he saw it. He also remembered specifically which door belonged to his mother's apartment in the building. After a little search he found himself in front of his mother's apartment. He knocked and a man answered him. Hitler told him who he was looking for and the man said he had never heard such a name before. Hitler was very disappointed and knew his only link to his mother was his great aunt but he didn't know her address but luckily, he remembered her full name.

When he was staying with his adopted father he was given the responsibility of going through his father's mails and reading them to his father. He remembered receiving annual mails from the municipality requesting his father to pay tax for his home. He knew that all homes were registered with the local municipalities. He didn't know if his great aunt owned the home she was living in or she was just a tenant. But he decided to try and found the office and gave them his great aunt's name. They sifted through the records for almost two hours and finally found her name and the corresponding address. He went straight to the address and felt lucky to find her. He saw that she had grown much older and gasped for breath every time she spoke because she had developed chronic asthma over the years. He told her why he came and wanted to know where his mother was.

His grandma wouldn't tell him where his mother was immediately and went on to narrate what happened since he was given up for adoption. She told him that his mother was a prostitute and thought it was the best for him to live in a stable home. Two weeks after she sent him to be adopted she couldn't live without him. So she went back to the parish to take him back but was told he was already taken by a family and would not release the identity or address of his adopted parents. She continued to return to the parish hoping that the parents who adopted him would return him because she was told that some parents returned children if they did not like them. She did so for two years and finally gave up. Later she settled with a man for five years but the man abused her badly that she had to divorce him. Suddenly she started to develop mental problem and freaked for no reason at all. She gave her shelter and lived with her for two years but then she became very ill and was admitted into psychiatric hospital. She was better after three months treatment and returned to live with her. After six months her condition was very bad and was readmitted into the hospital and that she has been there for the past two years. She also told him that his grandmother was a Jew and suffered from mental illness for many years of her life and died fully dejected and miserable.

Hitler got the address of the hospital and went to see his mother the next day. He was very sad but excited that he was going to meet his mother regardless of her condition. He met an elderly psychiatrist who identified himself as his mother's treating physician. He prepared him and told him that he may not see the mother he expected when he discovered that he had not seen her since the age of five. When Hitler came face to face with his birth mother he run to her, hugged her, and would not release her as he wept bitterly. His mother's hands were by her side and never responded equally. When they sat facing each other Hitler mainly spoke slowly and his mother uttered very few words. She said, "I know who you are. I know who you are" and Hitler was glad she recognized him for the doctor had told him that she was heavily sedated would likely not recognize him. Then she suddenly said, "I see the devil. I see the devil. I see the devil." Hitler didn't know what to make of it. But when she continued repeating it the attending nurse that was sitting in the room said, "It is very strange. She never said such a thing before. I think she is deteriorating." Later, the psychiatrist will tell him that it was not unusual for mental patients to see what they assume to be the devil, God, Jesus, angels and other spirits. Hitler was happy to see his mother's face which was still intact. She was older but the resemblance had not faded. She still had the smile he always craved for. She added a little weight, her hair was a little disheveled and wore hospital gown and other than that she was the mother he had always retained in his memory.

In the next two years Hitler will visit his mother four times. The second time he visited he noticed that his mother was a little thinner. The third time he came he was shocked. She looked so thin and wondered what was happening. He spoke to the psychiatrist and was told that his mother had just stopped eating. They had provided her with all kinds of food and even asked her what she desired to eat and whenever she suggested a particular type of food they brought it to her immediately. But still she just ate a little and rejected the rest. They kept food in her room for twenty-four hours hoping she would feel hungry and eat but had been unsuccessful. The doctor told him that her system just shut down for reasons they couldn't understand. They changed medications, reduced her dosage drastically but no sign of change. Regardless of her condition and appearance, however, Hitler still loved her abundantly and dreamed of living with her someday.

On the fourth visit to the hospital Hitler was taken to the psychiatrist first before seeing his mother. The doctor spoke to him for a few minutes and then revealed the bad news to him. He told him that his mother died three weeks earlier and was already buried. Hitler went ballistic. He sobbed loudly and tears rushed from his eyes. He went on and on without stopping for about ten minutes. The psychiatrist was very alarmed by the excessive anguish Hitler portrayed and knew it was very unusual. After all, Hitler knew that his mother was on the verge of death the last time he visited her and shouldn't have been surprised when he learnt that she was dead. So the doctor began to probe Hitler to discover what his real problem was. Hitler sobbed intermittently for the two hours he spent being interrogated by the doctor. At one instance the doctor asked Hitler, "what is your recollection of your great aunt with whom you lived for five years." Hitler became stiff and giddy suddenly and didn't utter a word. The doctor waited for a response and asked the same question again to Hitler when he did not get an answer for some time. Hitler just freaked and told him that he didn't want to talk about her and the doctor didn't insist. He doctor understood that Hitler did not have any affinity for his great aunt.

At the age of nineteen Hitler fell ill. He was nauseating and vomiting and his illness went on for ten days before he recovered. During that time he sensed strange feelings in his mind. He hated everything around him and felt very despondent. He had never felt that way before and only intermittent grudges against his great aunt. He felt disgusted with himself and fought the thoughts in his mind vigorously. His dejected mind felt better and soon he was normal again. He remembered that he has had such feelings before and was always successful in suppressing such thoughts. He continued his normal routines of the day and nothing changed after that experience. Hitler was an avid walker and strolled around the city virtually every evening. He always walked with faster strides compared to the other people that walked leisurely in the shopping centers of the city. He loved to watch people having a good time and conversing among each other. However, Hitler never mingled with them and was never a sociable person. He just had three acquaintances with whom he regularly met and spent some of his past time. Other than that Hitler was a loner.

One day, on his usual route, he stopped at the coffee shop he patronized. The coffee shop was always busy with clients and the tables were placed very close to each other to accommodate the many customers and one could hear the conversations going on adjacent tables. As he sipped his coffee he heard two men speaking and laughing loudly and wondered why they had to be a distraction. He couldn't help listening to their conversations because his table was next to his. He heard one of the men say, "our race is superior. We are the cream of mankind. No human being is comparable to us Germans. The Jews are inferior animals and have no place among Arians…" They went on and on and Hitler became interested in what they were saying and wanted to know more. He turned his seat towards them and began to inquire from them how they came to a conclusion after introducing himself and conversing for a little while. They told him that he should read about the evolution of man and how the Aryan race ended up as the most advanced and developed specie. He understood that there were books written on the subject. The next day he went to the local library and asked the librarian if they had books on the Aryan race. The librarian told him that she had never heard the name before and any book connected to it. She invited an elderly senior librarian and inquired from her but she also told Hitler that she was positive that there was no book on the topic he mentioned. He left disappointed and wondered if the two men were lying. Ten days later he recognized the two friends in the same coffee shop. He stood beside their table and told them of his experience at the library. They told him that he should search for books on human evolution or the evolution of man and he would find many books. He did so the next day and the librarian took him to a particular section of the library and he discovered that there were a total of twenty books on the subject. He picked five of them and walked to the front counter. However, he was told that he could borrow only two books at a time and so he left with only two. On subsequent days be would read all the twenty books and would become an expert in the evolution of mankind. Hitler was a highly intelligent person with an amazing memory. He could read a book and could write its replica just from memory.

Initially, Hitler never advanced beyond high school. He just never saw the importance of furthering his education. He gained admission in a university but dropped out in the first year. He was fully content with the vast knowledge he felt he had and never saw anybody as intelligent as he perceived he was. Hitler knew that he could succeed in any field he chose easily but he just didn't delight in anything. He never delighted in people and saw many of them as his inferiors and he truly had a point. His thought was highly refined and Hitler had the ability to analyze a situation to its barest minimum and suggest very viable remedies and solutions. He was just too good.

In his usual way, he continued his normal walk and his fascination with people even if he did not approach them. As usual, he passed by his regular coffee shop before he went home. He felt a little exhausted each time and had good sleep all the time. He never delighted in waking up early so he stayed in bed for most of the mornings. As usual, he made it a point to listen to the conversations of people sitting next his table and always wondered at the thoughts of various people. One day, he saw a Jew carrying many bags of items he bought from the shop he was exiting from. The Jew walked in front of him and finally joined a woman and two children who were waiting at the coffee shop where Hitler normally bought his coffee. He found the only available table which happened to be next to the Jewish family. He paid his attention to what they were saying and was surprised at the topic of their conversation. The woman seemed to be very angry when the man told her that he was not treated courteously when he purchased the goods from the shop. "I don't know what these evil people want us to do. No Jew hurt any white German but are all determined to harm us. To hell with all of them. I have no regard for any white German anymore..." said the woman angrily. The Jewish man said very little and only heard the woman vent her anger. Hitler never knew what to make of the woman's remarks and never took it seriously. Hitler never ever met a Jew nor had any conversation with any of them. He saw numerous of them on the streets but that was all about. He began to suspect that there was perhaps a strained relationship between Jews and white Germans. He remembered the two loud friends who continually denigrated the Jews also.

From then onward, he began to distinguish and identify Jews in his regular walks. He knew who they were mainly by their hats. He saw that they were peaceful people and never saw anything unusual about them. They were just like regular white Germans and wondered what the fuss was all about. He never saw anything inferior about them as the two loud friends claimed in their talks and didn't perceive any open animosity between white Germans and Jews. This went on for two years and Hitler never wavered from his normal routine. One day, as he walked in his usual fast stride he noticed that all other people walked slowly and clumsily. He began to think perhaps many of them suffered from mental problem. He remembered his mother at the hospital and how she did everything in slow motion due to the medication she was taking. He felt that that thought was absurd and discarded it from his mind quickly.

One day, when he woke up in the morning, he felt something unusual. His vision blurred and had a foggy sight. The symptoms persisted for most of the morning and then subsided gradually. But the next morning the same thing happened. After five days he went to an eye doctor to see if he was losing his eyesight. The doctor told him that he had a perfect eyesight after conducting some tests. And the fogginess never stopped. So he went to the hospital to be examined.

Hitler narrated his experience to the doctor and that he sometimes felt dizzy. The doctor finally informed him that he was having early brain damage symptoms. Hitler was shocked and went home in great despair. But after thirty days the whole fogginess went away and was normal again and Hitler thought that his brain had repaired itself.

Ten months later, Hitler began to see flashbacks and images of despicable circumstances. When the situation persisted he went back to the same doctor that had treated him earlier. Hitler told him that he saw brief images of gruesomely murdered men, dead women with their babies hanging partially born and attempting to get out of their wombs, headless children and other troubling images. The doctor told him that his brain damage was deteriorating and his brain was beginning to play tricks on him. Hitler believed him and was very despondent and felt that his life was almost coming to an end. He knew that his mother and grandmother died of mental illness and was almost certain that he inherited their genes.

Hitler's grandmother was a Jew. At the age of five she knew a handsome white German boy and they grew up together. Finally, they became lovers and married. However, his grandmother's Jewish family were totally opposed to their daughter marrying a white and Christian German and as a result, they disowned her and severed any form of contact with her. Her husband's white German family never accepted her fully and eventually ended up a sad and a lonely woman. She gave birth to a daughter and didn't have another child for the next ten years when Hitler's mother was born. Her husband was a flirt and had other women he patronized apart from her and she knew about them and he didn't hide it from her. He thought of divorcing him but she had nowhere to go and her family rejected her the few times she attempted to establish a link with them. By the time Hitler's mother was born, his grandmother was a depressed woman and a nervous wreck. She thought of committing suicide several times but never tried. Eventually her deep depression would result into a severe mental illness and she died a miserable and dejected woman. Hitler's grandmother never told her children that she was a Jew and did everything possible to integrate herself into the white community unsuccessfully.

Hitler's mother was very distraught when she gave up her son for adoption. She hoped that her son would be returned to her for two years and gave up finally. Eventually, she married a man that she met in the church where attended mass every Sunday. She was desperate to have another child. Her marriage was happy for the first ten months. One day, she made the mistake of accompanying her husband to a bar that she patronized as a prostitute. While they sat down a fat and ugly white German talked to her as if he knew her for a long time. He told her that he would wait for her, after she serviced the client she was with, to give him a sexual favor for he didn't know that he was her husband. The husband was very angry and quickly stood up and challenged the fat and ugly German for a fight. The fat and ugly man was surprised and tried to pacify her husband. He told him that they didn't have to fight over a "filthy prostitute" and revealed to him that he has had several encounters with her. The husband knew that the fat and ugly German was not lying. Later that evening he confronted his wife and she finally confessed and told him everything. He thought he had married a church going woman who helped in the church every Sunday and felt deceived when he learnt of her background. The next four years was full of torment and constant beating upon Hitler's mother. She became highly depressed for the husband never had relationship with her from the day he discovered her true history. Finally, Hitler's mother divorced him and kicked him out of her apartment.

She began to visit the bars to continue the only trade she knew but attracted very little attention from the men in the bars. She looked much older than her age and the clients preferred the younger women. She occasionally had a few clients and she spent most of the money on alcohol for she had begun to drink heavily and went home virtually drunk every night. She began to become more and more depressed and loathed everything about her. And she hardly had enough money to pay her rent and survive. Gradually, from chronic depression she degenerated and developed severe mental problem and began to freak, shout and became aggressive. That is when she began to live with her aunt and would ultimately die in a mental hospital. However, Hitler never had mental illness in the entire genealogy of his family. His grandmother and his birth mother developed mental problems due to the prevailing circumstances that afflicted them. Hitler, however, truly believed that he had history of mental illness in his family and feared that he would also be affected by their genes.

After the doctor told him that his brain damage was degenerating he became very distraught. One day, he went to his normal walk and sat down on the café later and began to ponder. He began to go through all the events of his life knowing that he was coming towards the end of it. He pondered and pondered and went through all the details of his past experience. Suddenly he realized what his problem was. He understood the level of impact his great aunt's actions had played in his life. He died a long time ago. He knew that he never grew above the age of five. All his later existence was just a farce and a sham. He really never grew up to be anything. His state of mind was totally governed by his tragic experience before the age of five. Hatred was overwhelming in his mind and he resented all human beings. His subsequent existence and actions were just mere pretence and the real Hitler was out of sight to every living being. Right then he knew that he was a sick man and a derailed human being. He fought his demons all these years and was able to exist as a normal human being. Hitler was his own best psychiatrist and devised various therapies for various symptoms of his psychosis and was very successful. He understood that he knew how to contain his illness and that was all it was.

He never ever had any ambition to be anything or become anybody. His main preoccupation was to just survive and live without any torment. The five-year-old Adolph knew how to survive and never died and he also did the same thing all his life. He saw himself as a man that was in the middle of the ocean and struggling to constantly keep his head afloat and survive. And that was all he was. Nothing more. He never ever saw himself as a member of his adopted parents and only lived with them to survive. Hatred never left his mind since the age of five and never knew love in his lifetime. His mother's love lasted for only six months and did little to quench the hatred he felt in his heart. Heaven and earth meant nothing to Hitler except his struggle for survival. Human feelings were devoid in his head and heart and knew no emotions. Hitler heavily relied on his intellect to survive. Everything he imagined was devilish right from the age of five but he really knew how to discard them. He kept himself sane and normal because he fought them vigorously. Humans know good from evil but Hitler never desired good in his heart and goodness was totally absent from his mind. From his perspectives all humans desired to hurt him and he constantly had to devise ways to coexist and to live with them and survive. That is how he saw the world. That is how much demented he was and Hitler knew precisely what he had become.

Immediately after that Hitler went into motion and was determined to change everything in the rest of his life. He immediately enrolled in an art school and began to pursue higher education. During his stay with his adopted parents he had enjoyed to draw and was very creative. He drew gruesome pictures mostly associating them with his wicked great aunt and his adopted father was highly impressed by what he saw and encouraged him so much. At the school he began to do the same and drew weird pictures which were mostly rejected by his tutors. He began to resent the field he had chosen and did not like the idea that he was painting pictures just to appease his lecturers and score high grades. He eventually graduated with the barest grade and he didn't care. All he knew was that he had a credential from an institute of higher learning. He searched for work but found no one that would hire him.

Hitler began to improve his social life also. On his walks he stopped by and conversed with a few people at a time. Gradually, he established many acquaintances who were happy to see him and enjoyed his grace. He became very talkative and just wouldn't stop when he started talking and all the people just listened to him. He spoke most of the time about his vast knowledge of the evolution of the human race and many were impressed by his intelligence. Soon, Hitler delighted in his newfound freedom and his interactions with numerous people. Hitler knew that he had overcome his illness and did not look back from then onward. His head cleared from devilish thoughts and avoided deadly thought from his memory. He had a hard time discarding them but he successfully suppressed them and was never tormented by his thoughts. He had no doubt that the thoughts of the tormented five-year-old child would continue to linger in his memory, however, he was no more enamored and captivated by his thoughts. Hitler became a brand new person and his hate was gradually dissipating and he never felt to disguise his true personality. Everything was perfect and gradually, Hitler became enamored with politics. He saw himself as a good candidate to vie for political offices. He met many candidates and learnt from them many valuable information and observed what they said to the people to gain their support. He knew that he was capable enough to do similar things and deliver powerful speeches and impress the people. Soon, he began to attend numerous meetings where vying candidates tabled their agendas and delivered speeches to explain them. And Hitler loved to see himself in their places and dreamed and started seeing himself as the center of attraction with many people listening attentively to him. He knew that he had finally found a profession that he felt he was born to pursue. He found his ultimate destiny.

As the time went on Hitler met so many other people and saw himself as the man of the people. His interaction was just amazing and his interaction with the people was also perfect. Knowing that he even imagined himself as the leader of Germany. The problem was that he didn't know on what platform he should run if he ever vied for a political position. He formulated many theories but found them not very viable. He figured out that not many Germans would support him for propagating the many ideas he had in his head. He thought and thought and couldn't come with a single message that he could use as his platform and run for an office in the local elections. He quickly abandoned the very thought of ever becoming a politician and reasoned with himself that it was a very long, long short and not worth the valuable time he would waste to pursue such a fruitless venture. However, he still went to political rallies and listened to numerous speeches. Nothing happened for two years.

One day, when Hitler was walking he saw one man hitting another innocent man on his head with a brick. When the man fell to the ground, the assailant sat on him and thoroughly bashed his head with his fist. After the aggressor was finished with the man on the ground he stood up and held a tooth that came out from the mouth of the beaten man and held it as a trophy and brandished it with a raised hand. The bloodied man attempted to rise but fell back repeatedly for it was very apparent that he was dazed. Eventually he succeeded and run for his life. Hitler was disgusted by the violence he saw. He observed and began to see the people around him and wondered why no one intervened to stop the one sided fight. He found out that they were all white Germans and none of them cared at what they observed for he realized that the person being assaulted was a Jew. There was no Jew around. Hitler shook his head and began to ponder and explored his thoughts and began his fact-findings then. It did not take him a long time to understand the primary reason why no white German came to the aid of the Jew. From then onward he noticed that basically all the white Germans expressed the same views of the Jews whenever he raised the topic among the white Germans. He quickly knew that the Jews were not Germans.

Hitler was always disguising himself as a full-blooded white German. He knew that his blood was mixed with Jewish blood and his uncanny thoughts he discarded that notion and never mentioned it to anyone. His thoughts towards the Jews, however, were never evil. He portrayed himself as a white Germans but was not particularly happy by the level white Germans resented the Jews. He really wanted to allay the fears of white Germans against the Jews and find ways to establish a real partnership between them. However, he discovered that the resentment of the Jews was very deep and no one could convince the white Germans to think otherwise. He delved deep into the history of the Jews and discovered that the Jews thought of themselves as the only authentic Israelites. Hitler never ever saw a bible in his lifetime and only attended churches when he was with his birthmother and up to the age of five-and-half years. He had no idea what Christianity stood for. But to understand the history of the Jews he read the bible thoroughly. He found that the word "Jew" was not mentioned frequently in the bible. Knowing that he began to contest the Jewish claims. He began to see himself and white Germans as the true Israelites. He really believed that the Jews were fabricating their history and raising their race above any other races, including white Germans. That infuriated him.

Soon, Hitler began to denigrate Jews with other white Germans. He even went further. He began to formulate his own theory from his vast knowledge of the evolution of man and successfully portrayed Jews as an inferior race. Hitler was a deadly person. He had the brainpower to convince anything to anybody and he knew his talent very well. Hitler had the secret of going into people's mind and knowing their sentiments shortly after meeting them. After all, he countered his thoughts and the perceptions of his mind throughout of his life and understood precisely how the human mind works. Having that vast knowledge he knew how to manipulate everyone he met. He was just too good. Hitler could use his smartness and eloquence to convince one that a chicken and a horse shared the same ancestry and used ninety-nine percent false information with one percent facts and still present it as the ultimate truth.

Hitler, not knowing, was drifting towards hating the Jews. However, he knew of any politician who addressed the concerns of the white Germans with regard to the Jews. He heard no message of disgust coming out of the mouth of any white German representing the people all over the country. He wondered why no one dared speak about them and against them. That will haunt him for a long time. He saw all white German representatives as fake and worthless and they did not listened to the voices and concerns of the people they were elected to represent. Still he did not do anything about it. It just infuriated him.

As Hitler advanced his name and preached the theory of the evolution of mankind he began to gain prominence. He was invited to attend a debate and speak on the theory. His opponent happened to be a Jew and the topic of debate was, "Does God exist." Hitler spoke at length explaining the theory of the evolution of man culminating into how the white German people ended up as the superior race with a highly developed brain and strong memory. He denigrated the Africans and the negro race as an inferior race with the least developed brain and virtually no memory power. He asserted that man eventually developed into apes and the presence of the negro race was a clear evidence to show a less developed brain that is a little higher than apes. He eventually concluded the non-existence of God and that the evolution of man had nothing to do with some supernatural being. When the Jewish opponent rose he obliterated Hitler's assertions. He quoted verses upon verses from the bible and somehow and successfully connected them with the development of mankind. He stated the separate developments of the humans all over the world and mentioned their achievements beginning from Israel to China, Native Indians, Europeans, Africans, etc. The Jews was just too good and Hitler was silenced, was lost of ideas and in total confusion. The Jew concluded that all the various regions of the world responded to the demands of their environments and had varied developments but were all precious people in the eyes of God. Hitler was given another chance to dispute the claims made by the Jewish opponent but only rumbled nonsense. The Jew rose up again and added even more facts.

At the end the panel asked questions and the contestants responded. The Jew went on and on and gave very convincing and credible responses. One of the panel members asked Hitler why there were no human species between the apes and the inferior negroes in the development of mankind. All Hitler could say was that the negro species became cannibals and ate the species that existed and were less developed than them and that is how they became extinct. They asked him if there were any archaeological findings to buttress his claims. All Hitler could say was that it was the conclusion reached by the superior brains of the Aryan race and their superior ability to comprehend, analyze and make credible deductions. At the end, the Jew won unanimously and received the price. Hitler felt that he was never humiliated like that day in his lifetime. He resented his Jewish opponent because he successfully defended his ancestry and equated them as equals to the white race. When Hitler described Jews as an inferior race, just higher than the negroes, his Jewish opponent gave himself as an example and challenged the panel to prove if there was an iota of evidence to show any difference between him and Hitler. He told them that he was probably even more intelligent than Hitler.

Hitler never knew anything about Jews until he was twenty-eight years old. From then onward he denigrated Jews and formulated various theories to discredit them. He often felt the desire to strangle them if he had the opportunity. The Jew who opposed him he felt put a dent in his reputation. Hitler never lost in an argument. He always came up on top. Hatred began to creep in Hitler's mind gradually. He delighted in the decisions of his mind and never tried to discard his hateful feelings. He loved the state of his mind because it gave him the vigor, vitality and aggressiveness he had noticed about himself and the new man he was evolving to be. Him and only mattered to him. Nothing else was important as long as he was able to ferment the desires of his mind. Healing was no longer a priority and even saw the tormented little child that still existed in him as a motivation force. Without the little tormented child he wouldn't have become the person he had finally become. Knowing that he began to tap into his memory and began to see the tormented child in his mind. He delighted in seeing him and did everything to console him from crying. Heavy surge of power went through his body every time he saw the little tormented child calmed down. He didn't know that he was degenerating into a full-blown psychotic individual. And Hitler never knew that he was a mental case in all his years to come and until his death.

Hatred was now engulfing Hitler. His head never thought of anything but murder, maiming, torture and other gruesome perceptions. In his early days he murdered his great aunt a million times and that is how he was able to cope with his stress. He mutilated her, burnt her alive, decapitated her, beheaded her, gouged her eyes, broke her limbs, slashed her throat and devised many deadly deeds in his mind. He felt safe every time he imposed his verdicts in his demented mind. Now, his symptoms were creeping gradually and were beginning to overwhelm him. Having power was all that was important to him and he did not care how he acquired such zeal. Hared he had known all his years, from infancy, and did not imagine any thing wrong with being obsessively hateful. Hatred was nothing new to Hitler, he lived it all his life. Having realized that he was hateful he began to vent venomous words through his mouth. His special targets were the Jews. He denigrated them with the strongest terms he could muster and made them very fearful of him. He met Jews and deleted them with hateful words but none of them hated him. They just thought he was a demented white German and ignored him in totality. The harassment continued until his deeds came to the attention of a wealthy Jew. The wealthy Jew delivered a very stern warning to Hitler and from then onward he became very fearful of the Jews. He never bothered them anymore but deep inside him he hated them with passion.

Very soon Hitler began to read messages from organized anti Jewish groups. He felt relieved and delighted that there were other white Germans who resented Jews like himself. He never met them and only read their pamphlets. One day, when he returned from his outing he found a leaflet stuck on the bars of the gate of his house. It was an invitation for anti Jewish rally. He looked at the venue and discovered that it was two blocks from his address. He quickly turned back and did not enter his house and headed straight to the address where the rally was being organized. He volunteered to help in the organization of the rally and was wholeheartedly accepted. At the end of the day, the group chose the main speakers at the rally and Hitler raised his hand and was chosen as one of the speakers. Ten minutes was slotted for each speaker. During the rally he realized that there were only about a hundred and fifty people and only fifty of them were close to the podium.

Hitler was the third to address the rally and he began mainly educating the audience about how the Germans became Aryans and a superior race. Then he began an onslaught upon the Jews and crushed to bits. He was so forceful, eloquent and delivered his speech with elegance. That is how good he was. He was slotted to speak for only ten minutes but he went on for more than half and hour before he was stopped by the leader of the rally. He got a huge applause and his ego almost burst. Nothing happened for three days and when he glanced at the papers on the newsstand he saw his picture on the front page of a tiny newspaper in the city. He bought the newspaper and read it and most of what he said in the rally was well reported. His imagination began to run amuck. He hailed himself for the speech he gave and was so proud of himself.

Subsequent days followed he began to leave his home to attend rallies against Jews. Soon, he was invited to most of the rallies and asked to give speeches, which he did with the usual veracity. Two years later he would become the main guest speaker and to much larger audiences. He became a deadly opponent of the Jews and the Jews took note of him. Newspaper articles began to appear totally discrediting Hitler and his bogus theory of the evolution of the Aryan race and he knew they were Jewish newspapers. The more such articles appeared the more Hitler became popular among the radical segment of the white German society. He began to propagate a party and named it Nazi after a very ancient tribe in Germany that he read in a deadly, evil, distorted book by a German philosopher. Many people began to identify themselves with his party but never gained popularity among the overwhelming majority of the German nation.

He wanted to call his party a different name when he did not attain the required popularity but kicked against it. His head turned upside down and he began to disappear in the midst of the turmoil that existed in successive German governments. Messages of the German leaders were mostly empty promises and the people relegated them as liars and had no confidence in any o them. Chaos was the order of the day at that time and there was no stable government in Germany. Frantic politicians made numerous and impossible promises each advocating a very viable and prosperous Germany. None of them delivered their promises and Germany remained still shambled economically. Hitler understood the exact situation of the nation and began to formulate hateful messages blaming the Jews for the near demise of Germany. And his message resonated among the people. Finally, the Germans began to understand why their nation was not developing and falsely found scapegoats. The Jews were identified as owners of a huge chunk of Germany. Every Jewish asset was considered as detriment to the existence and survival of the German nation and discarding them was the only solution. Hitler bade them to elect him and promised to deliver every agenda of his party to the last letter.

His grave hatred towards Jews became a great concern to the German government. Even the pope in the Vatican was alarmed by his veracity. The German government dealt a big blow to Hitler when they trumped up charges and accused him of sedation. He went to trial and was sentenced to two years imprisonment but got out after one year. The pope began to monitor specifically the activities of Hitler and was briefed regularly because he saw Hitler as a threat to the Catholic Church. Hitler never identified himself as a Catholic and was not in favor of the Catholic dogma. Unfortunately, the pope will be two years too late when he asked his flocks in Germany to declare war on Hitler and his evil philosophy. Ultimately, not even the powerful Catholic Church could put a dent in Hitler's popularity and will become a victim of his evilness.

Hitler propagated his dogma unopposed and spread his deadly ideas fearlessly. He spoke in rallies, he spoke at conventions, he spoke in schools and universities and in town halls. He was everywhere. He never wavered from his vision of becoming the German Chancellor. At the end, no one could stand against him. He never wavered from his stance to discard the Jews from the German society. Not only that, he knew precisely how to get rid of them but he kept the secret to himself. Evilness was his mantra and knowing that he desired nothing but the extermination of the Jews. In his evil perception, Jewish were no better than locusts and they swarmed a society and deleted everything that was viable. He believed that no heavenly bodies have any solutions as to how to discard the Jews. He was the only mind in the entire world that had the capacity to implement the final solution to the very existence of the Jews.

Ultimately, Hitler would delete all his opponents and reduce them to nonentities. One day, when he was watching the news on television he heard a prominent and promising German politician say, "the Jews are our blood brothers. Germany will not exist without the Jews…" Hitler shook his head with amazement. Right there and then he knew that he was the next German Chancellor and he never wavered from that conviction until he finally attained the ultimate power in the German government.

Hitler never knew from what section of Germany he belonged to. He merely knew that he was half white German and half Jew. He never deceived himself and Hitler never indulged in self-deception. He was severely demented, however, he never lost his sense of reality. Hitler hated the Jews not because they were destructive and delivered heavy blows to white Germans and denigrated them but for an entirely different reasons that no other white German understood. Hitler had a very highly developed power of thought. He made every effort and spent days pondering to know the core sources of many events and issues and his thoughts were not superficial. His knowledge to analyze and breakdown information to its barest component and then rebuild them again to arrive at a prevailing situation became his obsession. Hitler analyzed his mind consistently and recommended solutions that he carried out effectively throughout his life. His head was a thinking machine and never delighted if his thoughts were not engaged in one thing or the other. Hitler's hatred towards the Jews had nothing to do with their faith. The reason of his hatred he never told anybody for that was his motivational power and didn't want to reveal it to his political opponents. He merely propagated white German supremacy and designated the Jews as inferiors and knew that it was the appropriated message that resonated among the majority of the white German population. His facts, however, were totally different and Hitler knew them. The truth about the Jews was far more sinister and dangerous to the very existence of the white German society and the German nation.

Hitler never deleted the Jews as a way to pump his ego or to proclaim a white supremacist Germany. Over the years, Hitler spent countless hours in deep thoughts and wanted to understand fully why white Germans resented the Jews. He read extensively the entire German history and finally was delighted to pinpoint the actual source of the animosity that existed in the hearts of the overwhelming people of Germany. Germans did not just wake up one day and began to hate the Jews. He discovered that then prevailing situation was instituted in the minds of Germans over many centuries. He knew exactly how it started.

He read the entire history of Germany and knew that the Germans were a very ferocious, brave, valiant and sometimes even destructive people in Europe. Their entire history was dotted with extreme rivalry heavy wars and deadly marauders. The Germans were a bunch of wild people. But that stopped suddenly and Germany was inhabited by a docile and base human beings that had no territorial ambitions and lived as clowns. He wondered what must have taken place to transform them from a set of wild people and end up as ambitionless, meek and worthless people. He pointed out and identified the change with the migration of the Jews to Germany. The realized that the Jews were experts in mind control. They were never violent but were deadly in dominating a people without them knowing it. He associated them with a snake that coils and pretends to be harmless but has a deadly sting. He understood the methods of the Jews and how they stealthily influenced the minds of the people they lived in. Never fought physical wars but were in constant battle against the German people and put fears and terrors in the minds of the white Germans.

No influential German dared speak against a single Jew because Germans became mere stooges of the Jews and had no mind of their own over three and half centuries. Hitler wondered how they did it. He noticed the ultimate Jewish unity. He went on to read the Torah and found in the library over five hundred laws by which they Jews were governed. He realized that they were merely a dangerous and deadly cult that dealt heavy blows on the society they lived in and basing their origin to the Old Testament. He knew also that Jews never wavered from the cult constitution that governed their lives. Hatred was in the Jewish minds, even if it was not manifested openly, but their ultimate agenda was to bring down the German people to their knees and they did it successfully over the past three centuries. Germans were transformed to worthless goons by the Jews and were dead people. As a result, Hitler saw no German worth his respect. He ultimately saw the German people as a herd of cattle that needed a shepherd and that sole shepherd was him. Hitler had the ultimate plan for the Jews and getting rid of them from the midst of the German society became his primary priority. He was never sure if he could ever revive the old German pride, nationalism and veracity but the Jews had to go and there was no compromise there.

Hitler never hated anybody but developed extreme animosity towards anybody who challenged his messages that he felt were the ultimate truth. He knew that his conclusions were based on incessant thoughts and very rational analysis. He never saw any human being that could be compared to him and he had a point. Never was a man with such a powerful mind and brain that could figure out facts like Hitler. If Hitler had chosen to pursue science he would have become one of the most influential scientists and could have had many inventions and theories to his name. But that was not meant to be and he became a deadly and disgusting figure instead. He had a pumped up ego like no other man before him. He idolized evilness instead of goodness primarily because of his tragic childhood experience. Hitler never knew goodness in his mind and was always obsessed with vengeance. Never did he forgive those who stood on his way and always thought of deadly retributions against them. Ultimately, hundreds of thousands of white German citizens would die for merely opposing his evil ways. No one knew the real Hitler and he was very good at portraying himself only as the fake Hitler the people would idolize. If anyone knew the real Hitler he would have been traumatized for life.

Hitler understood the methods of the Jews and their sinister and deadly ways of brainwashing those they lived with and reduced them to their subjects. Hatred was their primary motive. Hitler knew that Jews were a hateful bunch of people who inflicted their vengeance with great articulation. Hitler, knowing that, decided to pay them in their own coins. They inflicted hatred towards Germans and deserved nothing but hatred in return. Hitler fought the demons in his mind consistently for many years but when he figured out the final solution to the problems of the Jews in Germany he became excited and never fought his devilish idea. He knew that burning them alive was the ultimate solution as a vengeance for the reduction of the German society to mere zombies over three centuries. And he did just that.

Ultimately, Hitler fried Jews in ovens and deleted a total of just over two hundred thousands of them. The Jewish hatred disappeared in Germany and Hitler revived German pride and nationalism and built a truly independent German nation. Germans hailed their identities and grew in the knowledge of themselves. For over three hundred years they had accepted the Jews as the only chosen people of God that the Jews falsely propagated and were relegated to second-class citizens in their own country. Hitler became their God sent liberator and he indeed was. God heard the prayers and faithfulness of the German people and the good Lord instructed Satan to do what he best does and to inflict pain upon the Jewish people. Hitler became the best candidate for Satan to carry out his devilish design. Hitler was possessed by Satan from his early childhood but he fought him for most of his life but gave in at the end. Satan won. But regardless of the crude method used to exterminate the Jews, Hitler was created to liberate the German people from mental bondage and slavery. And that was the ultimate design of God and the German people were finally liberated from the Jews and have now a mind of their own and great pride in their nation. They paid dearly for following Hitler blindly but at the end it was worth it. If it wasn't for Hitler the German people would have disappeared into oblivion and the Jewish people would have dominated them wholly.

During the war, Hitler got regular updates and knew precisely how many Jews were gassed, fried, shot and died of hunger. Hitler never flinched for a millionth of a second and understood that his ultimate verdict and sentencing of the Jews was taking place effectively. His heart was done and had no human conscience. He saw the image of his tormented childhood throughout all his life but was finally successful in pacify the child. Every time the child was quiet he felt that he was doing the right thing. Hatred was not the mantra of Hitler but the obsession for deadly vengeance became a component of his existence.

Vengeance, vengeance and vengeance was all he thought about Hitler didn't know that his great aunt died barely two months after he last saw her when she gasped for air, tuned blue and expired. He always thought she was alive and never found out the truth. Perhaps, if he had known of her death his mind would have relegated his obsession with vengeance and would have become a different person. No one could tell. Hitler always regretted for not striking his great aunt on the head, mutilated her and burnt her body in the fireplace when he last saw her.

No one could also tell if he had indeed murdered his great aunt his destiny would have changed a bit. Hitler was already a damaged good by the age of five year and half. He was a dead man walking. Hitler learnt how to survive at a very early age and became an expert in the art and techniques of survival. He was just a survivalist who lost basic human conscience in his early childhood. He was the man in the middle of the vast ocean who succeeded to keep his head above the water throughout his life. Nothing more. He didn't care how many people drowned and fell to his left and to his right but his only survival was his ultimate priority and that was what he was governed with until his death.

At the early age of his life Hitler never maimed animals and never gave an impression of his bogus personality mainly because he did not accept the thoughts that constantly crept in his mind. Heavy and powerful mind delivered him out of the psychotic symptoms he was beleaguered with in all his life. At the age of twenty-nine he lost that ability and became a victim of his thoughts and could no longer contain his psychopathic tendencies. However, he was good at covering up the real messages that bombarded his mind constantly. Not many people with mental problems could do that. Hitler never delighted in great messages but delved in trivial and simple messages that were understandable to the German people. He never made things complex and delivered great speeches transmitting simple and specific messages. And all the speeches he made were absorbed by the German people wholly. That was his talent and stooped very low to the basic understanding of the commonest man in Germany. He knew how to deliver a very complicated message in the simplest way available.

At the end Hitler had the desire to exterminate anyone that sought his destruction. Hitler never wavered from such a desire and became his guiding principle. No ones message was relevant to him and depended entirely on his own decision. Hitler made amazing decisions in the early years of his leadership and revamped the German economy. At last, he started the second world war and delighted in his achievements and uplifting the greatness of Germany. He made crucial decisions and invaded neighboring nations. Hitler could have occupied the entire Western Europe, retained and maintained it.

No British and American power could have dislodged him and he knew it. But he made the gravest error of his life when he invaded Russia. That would spell the doom of his life and an end to his ambition to unite Europe. His mind was not consulted when he ordered an attack on the Russians. If he had, he would have had the conclusion that it was not a viable thing to do. Death became very rampant and heavy casualties were inflicted on the German army constantly. Hitler's head was not normal in the last two years of his existence. His head and mind were foggy and didn't have the rest he required.

He became just a ruthless human being with no power of his mind intact. He suffered from paranoia and was very delusional. He didn't know where he was heading to any longer. He never wavered from the extermination of the Jews and showed no emotion by the massive German lives lost. Until the end, Hitler thought he was winning the war and never believed that he would fail. That was how much screwed up he was. Hitler would ultimately commit suicide when he saw no other way out and that would end the history of Hitler.

When the five thousand Jews landed on the shores of Portugal they were welcomed with mixed messages. The northern Portuguese were furious when the Roman brought new immigrants and were determined to exterminate them. But a small group of people in the southern tip of Portugal thought differently and harbored them and saw them as allies. These group of people were originally from Mesopotamia, present day Armenia, and migrated to the region shortly before the Roman conquest of Portugal. They left Mesopotamia because the region they came from was inflicted by a plague. Dangerous metals began to wash into their rivers when the inhabitants ate the fish from the rivers they developed chronic liver damages and began to die in large numbers.

Neighboring villages that did not eat fish from the poisoned rivers were fearful and knew that it was just a matter of time before they were also inflicted with the same plague. Hence, thousands of them began to migrate and twenty thousand of them found themselves in southern Portugal. The Portuguese inhabitants fought against them but were quickly subdued by the superior weapons of the Mesopotamians. The Mesopotamians pursued the Portuguese for many kilometers and occupied the land they conquered as their own and usurped numerous women from the locals and established themselves fully. The Mesopotamians always had kings and hence, the new settlers anointed a leader as their king and started the monarchy in Portugal. The northerners never accepted the Mesopotamians as part of their people and for a long time never recognized the monarchy.

The monarchy helped the Jews integrate and relied on them for support. As the years passed the bond between the monarchy and the Jewish people will solidify. The Jews will learn the history of their new home and will identify their friends and their enemies. They will multiply and flourish in the southern part of Portugal and never dared advance to the north. The north was a fully Christian region and the south will become partly Jewish.

When the Jews were finally driven out of Portugal there was a proclamation from the monarchy prohibiting any Portuguese from acquiring their gold and silvers. Those found in possession were hanged by the monarchy and two hundred of them went to the gallows at the end. All the wealth of the Jews were handed over to the monarchy and it was mind bugling. When the soldiers entered into the homes of numerous wealthy Jews what they saw shocked them. They had bails and bails of sacs filled with gold and silver coins. Some of the sacs were so heavy that two people had to carry them to the awaiting carts. At the end the monarchy had no place to store them and had to construct special vaults for the vast Jewish wealth. They didn't count them but they were in the hundreds of thousands. Suddenly, the central government was so rich that it never solicited taxes for the next twenty years. The homes of the Jews were taken by whoever occupied them first and the overwhelming majority of the Portuguese people in the south became home and landowners.

The monarchy was immediately approached by some ship builders and requested to be funded to build a mighty ship that could withstand the huge tidal waves of the ocean. The monarchy complied and a ship was built within two years. The ship set sail and did not return until after two years. The sailors brought good news and explained to the monarchy the various lands they visited and including Asia. They told the monarchy that there was so much gold in some of the lands they visited and that their women were covered virtually with gold ornaments. The monarchy was encouraged and ordered the building of ten more mighty ships and they had the Jewish wealth to fund the project. More than ten thousand people will work on the ten ships and will be completed in three years.

Meanwhile, the first ship will sail once again to discover more territories and this time had many soldiers to defend against hostile natives of the lands. That ship will never be seen again and it never returned to Portugal. However, the other ships will sail towards the Americas and will find abundant gold in present day Brazil. As a result, Portugal will prosper in wealth and become a rich nation.

The successful Portuguese adventures reached the king of Spain after ten years. The Spanish had the knowledge to build ships but did not have the resources to construct huge ships until the Jews were departed. As in Portugal the king of Spain had become extremely rich also thanks to the vast wealth left by the deported Jews. He immediately ordered the construction of five ships and were completed in two years. They immediately set sail but were unlucky and did not find lands with abundance of gold. But still the king continued to build more mighty ships and at the end had thirty-six of them sailing to colonize as many colonies as possible. Spanish ships were loaded with cannons and armed with experienced soldiers and they brutalized the local inhabitants they found into submission. But still they could not find enough to sustain the monarchy in Spain for it had exhausted the wealth they accumulated from the Jews and the Spanish had not paid taxes for almost twenty years. When the king of Spain attempted to collect taxes from the people again they all refused. The king went to war followed by his heavily armed soldiers. The king would finally die in a bloody war with the local people and the monarchy will never again regain the supreme power it wielded previously.

The British will send fifty of its most talented ship builder to Portugal to learn the art of building mighty ships. After two years of experience they would return to Britain and within fifty years Britain will build more than two hundred ships. The French will follow suit also. But Germany will not become a ship builder and never explored the seas. While tiny nations like Portugal and Spain conquered and colonized many territories Germany will be totally left out from the competition. Hitler felt it was because of the presence of the Jews who crippled the minds of the German people and made them less ambitious. He had no credible evidence but regardless he truly believed it to be the ultimate truth.

When the Jews initially began to flow into Germany they didn't identify themselves as Jews nor told them where they came from and certainly did not reveal that the pope was responsible for their deportation. Gradually, however, they became more comfortable in Germany to reveal their true identities as Jews but formulated their own story and said that they migrated from Jerusalem instead of Portugal. The Jews would propagate this story and not even their subsequent generation would know that they were deported from Portugal and Spain and believe that they originated from Jerusalem and migrated to Germany. All the Jews had changed their names and adopted German names and did not tell their children that they were deported from Portugal and Spain because they didn't want to burden then and impart an inferior feeling among them and to see themselves as rejects and dejects.

Also, there was no record in the history of Germany to show that there was massive migration of Jews into Germany. It was not a milestone in the history of Germany. The four hundred thousand Jews were hired as farmhands and were scattered throughout Germany and melted into the population immediately. There was no concentration of Jews anywhere in Germany. At last, no one would know, Jews and Germans, precisely where the Jews migrated from and when.

The Jews came with the idea and began to promote that they came to Germany during the Roman dispersion in 70 AD, and the Germans believed them. Hitler did not know also when the Jews came to Germany and from where they migrated. He bought into the Jews falsified history. But one day when Hitler was glancing through the titles of the books in a library he saw a book with an old binding. He began to skim through it and found out that it revealed many important events in the history of Germany. He took it home and read it. Towards the end of the pages he found just one page on the Jews.

The book revealed that all the Jews were deported from Portugal and settled in Germany and the exact year they entered into Germany. The book gave the estimated number of Jews that came to Germany as two hundred thousand. Hitler instantly assumed and believed that the Jews must have committed something evil for so many of them to be deported from Portugal and that will further enhance his stance against the Jews and his determination to stamp them out of the German society. He then knew why German history went silent suddenly and coincided with the era of Jewish migration into Germany. To Hitler, his puzzle was almost complete and so also his resolve.

Before the deportation of the Jews from Spain and Portugal there were only fifty thousand of then in the entire continent of Europe. More than ninety percent of them migrated during the Roman Empire and were primarily skilled artisans and found it more profitable to work in Europe. These Jews lived in peace with their surroundings even though they were sometimes persecuted by the church and accused of involving themselves in witchcraft and other anti Christian activities. However they lived and survived as normal citizens of the various countries they inhabited for many centuries.

When the deported Jews arrived into Germany there were only two thousand of them in the country and were all landowners and fairly wealthy. These Jews had no Torah and did not have vast knowledge of Judaism but observed the Sabbath throughout their history. When they learnt of the new Jewish immigrants they were very excited. They brought two hundred of them to Hanover, where they were concentrated, and became avid students of the Torah and leant Hebrew also. Eventually, they would be the main instruments to influence the Jews working as farmhands to go to the towns and cities and abandon the exploitative farmers.

And virtually every Jew did eventually. Most of these ancient Germans would become fanatic Jews and will ultimately return to Jerusalem shortly thereafter. However, Hanover would become the center of Jewish learning and their Vatican with the head rabbi like the pope giving instructions to all Jews for many years and until the advent of Hitler. So when the rabbis formulated the theory that they migrated to Germany during the Roman Empire they had a point and the Germans knew that they always had Jews among them for many centuries and hence, accepted them as part of their society all the years they stayed in Germany.

When the Jews migrated from Portugal most of them settled in Germany where they lived and prospered until the advent of Hitler. Many Western Europeans dealt heavy blows on them and many of them were dead. Out of the Portuguese Jews twenty thousand of them found their ways back to Jerusalem, almost all the rabbis. They had twenty Torahs with them and were well prepared for their new settlement. When they got to Palestine they were shocked not to find a single Jew in the entire country. Initially, they thought they were going to be exterminated by the Moslem Arabs but were surprised at their kindness and hospitality.

The Palestinians understood that the Jews were a lost people of Israel for they knew that their land was the land of Israel. Hence, no Palestinian ever saw the Jews as outsiders and accepted them as one of their own. Gradually, news began to spread all over Europe that Jews could return to Palestine and live in harmony with the Moslem Palestinians. In over three hundred years the Jewish population in Palestine will swell to over five hundred thousand people. Still, the Palestinians did not express any concern and were not worried. However, the Jews began to resent the Moslems and began to see Palestine as their own land of inheritance.

Having the required population the Jews began their determination to get rid of the indigenes and replace them with their own members. The Palestinians did not show much resistance and delivered their homes and lands to the Jews for very little fees. Most of the original indigenes were glad to go somewhere else because of the incessant threats, intimidations and harassments by the Jews. Gradually most parts of Western Jerusalem fell into the hands of the devil Jews. However, as time went on, the Jews even went much further.

They did no more buy the properties of the Palestinian indigenes but drove them out of their lands forcefully and confiscated their properties. The original indigenes were highly disgusted and made many efforts to fight back. The British protectorate that administered Palestine were very sympathetic with the indigenous Palestinians and they intervened on their behalf. They passed laws prohibiting Jews from forcefully confiscating the original indigenous' lands and properties. The devils, however, wouldn't heed and observe the dictates of the law passed by their British colonial masters.

Hitler wiped out two hundred thousands of the two million Jews that inhabited Germany for nearly four hundred years. Out of the surviving Jews two million of them eventually migrated to Palestine and mingled among the original indigenous Israelites that had lived in the region since the time of Abraham. The German immigrants would ultimately be a nightmare to the indigenous Israelites. The British made frantic efforts to dissuade the wicked and evil German Jews and made many efforts to establish some sort of reconciliation between Satan's children and the hapless indigenous Israelites, the Palestinian people. The British failed woefully and the German Satan's children ultimately drove them out of Palestine. The devils made a proclamation and anointed themselves as the rightful heirs to the land of Palestine and declared themselves as the sole and only authentic Israelites. They made demands to virtually annihilate the rightful owners of the land and true and authentic Israelites and declared themselves as the chosen people of God. In 1948 Satan's only nation in the world was created. And the message went throughout the world and mainly wicked nations recognized them. And they named the devil's nation Israel.

No one knew that evilness would engulf the world with the creation of Satan's domain called the state of Israel. The sickly demented German Jews would have no problem consolidating their grasp on the entire region of Palestine. Heavy messages will be dispersed by the sickly devils among them and will manifest their deadly plans against the innocent, helpless and dispersed authentic children of Israel. Their rabid dogs anointed themselves as rabbis and experts on Old Testament records. Heavy reliance will ultimately be made upon them by many disoriented Christians and they will deliver utterly false interpretations of the Old Testament. Their mind is that of the devil and their messages also distorted desires of demented humans. Ultimately, no one will dispute the claims of the devil German Jews and their names will become synonymous with the original state of Israel and would safely be known as the only chosen children of God. And the heart of the Lord ached.

The German immigrants will not confine their wickedness to the creation of Israel, alone. They would desire to maim and destroy any nation that stood against them. The main brunt of their hatred would be the African continent. Israel would declare war on the impoverished and helpless continent of Africa. Virtually every man in Africa would eventually realize the enormous powers of Israel. They delved into every African nations' affairs and nearly chose who would lead, and how, of the various countries in Africa. Any African nation that did not tow their lines was punished with impunity. Their deadly terrorist secret service, the Mossad, became the nightmare of the continent. Israel saw the African continent as an extension of their territory and saw the African people as mere disposable commodities.

In the sixtieth, seventieth and eightieth Israel hailed disgusting and mad dictators and supported them with arms supply and logistics to oppress and dominate the African people. The continent saw no noticeable development during that period of time and Israel did not care. No one in Africa ever maimed or tortured any Jew, however, millions of Africans were maimed and murdered as a result of the Israel's intervention in the continent.

In Angola, Ronald Reagan, the American president, utilize the expertise of the state of Israel to inflict massive casualties in the country. As a result, over two hundred thousand innocent Angolans were murdered in the civil war that was triggered and led by the state of Israel. In Liberia and Sierra Leone, in West Africa, many, many Africans were amputated, raped and murdered by despicable warlords that were exclusively trained, encouraged and armed by the Israelis at the beginning.

The war started when the then Liberian leader, Samuel Doe, who was placed in power by the state of Israel, betrayed them and visited Gaddafi and declared his support for the Palestinian people. That was enough to trigger and institute deadly revenge that will cost the region heavily and leave hundreds of thousands of dead Africans.

Apartheid in South Africa was sustained mainly by the flagrant support of the white supremacists by the nation of Israel. They never made their unflinching support a secret and armed and gave logistical support to the apartheid regime and in total disdain of the African people and the black race. In Eritrea, the Jews armed and trained a very deadly forces called, "commandos" during the Eritrean war for liberation.

It is very apparent that Africans will not be able to decide their destinies independently and see any credible development in the continent as long state of Israel exists. Even though most of the African nations are today asserting their powers and attempting to decide their ultimate destinies they will always remain under the radars of the Jewish state. Africans are terrorized by Mossad and fear the organization more than God.

In the 1985 draught and famine that ravaged the entire northern Ethiopia, every country in the world was alarmed and help came from everywhere to alleviate the sufferings of the people. Hundreds of thousands died before help could come to them because the sickly inhuman Emperor Haile Silassie tried to cover it up and refused to acknowledge that a problem existed in his northern region. At the time the Jews engaged in evacuating just those they felt were Jews like themselves, the Falashas, and transported them to their state of Israel. They assume that the Falashas, the Beta Israel, have Jewish origin. Nothing can be further from the truth. The Falashas are the direct descendants of Manasseh, son of Joseph, son of Jacob and have no direct relationship with the Babylonian Jews. The Falashas have eighty percent Israeli blood while the Jews have only twenty percent of Israeli blood.

Hitler was never desiring to destroy them initially but merely to send them to Palestine from where he assumed they came from. His final deadly solution was arrived when they warred against him and dealt heavy blows to his messages. Their cunning way of making their voices heard impressed him so much. Hitler finally did what he did because he realized that the Jews were a very deadly specie of the human race, from his perspectives. He deleted them finally because he felt that he was doing humanity a favor by exterminating them.

As memory could remember human species evolved from the breath of God. Had no one became human we would have no one to praise the Lord and glorify his name. Many centuries ago God decided to select a very small number of people that were descendants of Abraham. They called themselves Israelites. Hatred was not in their minds and were precious in the eyes of God and the Lord protected them and multiplied them. He swore to Abraham, their forefather, to grant the entire earth to his descendants and their numbers would be as many as the stars in heaven. Eventually, the Israelites would disintegrate and some would go to Europe and others to Africa. The ten tribes of Israel would be Europeans and only one tribe, Manasseh, would be Africans. The tribes of Judah, Benjamin and Levites would integrate with the descendants of the ten tribes of Israel and all the thirteen tribes of Israel would be known as the remnants of the house of Israel and will become the Palestinian people.

God will delete every other human and replace them by the descendants of Abraham in the end. Now, we have no living human being that is not the descendant of Abraham except the Jews. Eventually hatred will disappear and the children of God shall live in harmony in the whole wide world. Deadly wickedness will be no more and no one will desire evilness against the other. Man shall be a lively specie and many shall believe in the Lord and his blessed Son, Jesus Christ. Creation shall be known and the theory of the evolution of man shall be discarded. Humans shall give their lives to protect one another. Joy and happiness shall prevail all over the world. There shall be no blacks, whites, Asians, Arabs etc. and shall be only one humanity.

Man shall never deviate from the Lord's way for the next two thousand and five hundred years. After that, human species will destroy one another and eighty percent of humanity shall perish. Man shall devise deadly and lethal weapons that shall render the present nuclear weapons a mere child's play. The Son of God shall descend from the heavens then and pacify the human specie. And at that time, God shall begin to manifest himself to mankind as in the days of Adam and reveal himself unto mankind.

No one shall doubt the existence of God. And that shall continue for the next two billion years. Man shall have the ability to replenish his body and shall live ten folds than its present age. After two billion years the human specie shall plot to overthrow the power of God and as a result, the Lord shall destroy the earth with a single word just as he created it.

There are no living things in the entire universe. No man has the brain capacity to really imagine the expanse of the universe and it is limitless and endless. After the destruction of the earth, the Lord shall create new species but none of them shall be in the image of God, like the human species. Hundreds of planets shall have living beings in them after the end of the earth. Many events shall occur between now and the next two thousand five hundred years. Many nations shall rise and many nations shall fall and there will be no third world war until after two thousand five hundred years. Humans shall know no bitterness and their minds shall pacify and shall desire no evilness among themselves. However, the world shall be reshaped constantly. Knowledge will be expanded and daring things will occur that will defy human imagination. Health will improve and man would live to two hundred years in the next one thousand years.

The continent of Africa will be the most affected and will see total transformation in the next two thousand years. If Africans don't plan seriously and address the cutting of trees the continent shall become total desert before the next two thousand years. No humans shall live in it and shall be composed of just sand dunes and with no living things existing in the entire continent. The rivers and lakes shall dry up and there will be no rain in the continent. Black people will exist only in other continents.

If Africans don't stop the cutting of trees the total desertification of the continent could take place much earlier. Regardless, however, the African continent will be fully desert after two thousand years. Knowing that the entire world should immediately take action and not leave the environmental policies under the discretion of the African people alone. Policies should be made universally to delay the desertification of Africa and the preservation of humans and animals in the continent. Africa will blossom to four billion people in the next three hundred years and every inch of Africa will be occupied by people. If that is allowed to proceed a third of arable land in Africa will be desert in the next seven hundred years. The next one hundred years more and more land will disappear to the desert encroachment. And in not too distant the whole continent will disappear and become totally sand dunes.

The whole world has a responsibility to become involved and restrict Africans from pursuing a destructive path to oblivion. No one should tolerate the disappearance of a huge chunk of arable land of the earth because of the silliness of the African people. Pressure should be made on African people not to reproduce at the alarming rate they show now. Not one should have more that three children in the continent to sustain life in Africa. Not only that pressure should be mounted on the continent to totally stop the cutting of trees and instead plant more tree. That could only be achieved with the involvement of the entire world. However, the destiny of Africa could be reversed if the inhabitants could change their evil, wicked, filthy and despicable ways and absolutely give their hearts and souls to the Lord and his blessed son, Jesus Christ, for there is nothing the Lord cannot change and restore. If Africans would come out of their witchcraft, ancestral and devil worship and heathenness and wholly submit themselves to the ways and blessings of the Lord and Jesus Christ even the Sahara and Kalahari deserts will blossom and become farmable and Africa will sustain life forever. The choice is entirely in the hands of the African people and the destiny they choose to pursue. Death or life.

Africa will become one nation ruled by one leader immediately. Africans will know their ancestry and understand that they are the precious children of Israel, from the tribe of Manasseh. The African Bible, Book of Mormon, will authenticate their rightful identity. They will know that all Africans are blood brothers regardless of what country they belonged to. Colonial boundaries will disappear and no Africa will require any document to travel anywhere in the continent. Heavy burden will be placed on the knowledge of God and they shall be motivated by their discovery that they are authentic children of Israel.

Great things will happen in Africa when they eventually learn of their true ancestry. Hatred will disappear and tribal differences will be minimal and insignificant. Africa will be united like no other nation before it. Massive investment will take place in the continent when other nations realize that they are their brothers. Africa will no more be exclusively black and many other races will migrate to the continent and will be accepted by the local indigenes. Everyone is a descendant of Abraham in the world. Things will change very much in Africa but the entire world will come to its aid and relieve it from imminent extinction.

In exactly two hundred and eighty-eight years Australia will be engulfed by the ocean and ninety percent of its land will be covered by water. The ten percent that will remain afloat will be the desert part of the country and as a result, there shall be no living beings or animals in Australia. Again, if the Australian people anoint the Lord and His son, Jesus Christ, as the Lord, master and king, as Africans should, their tragic destination could be reversed. Instead of total destruction in the Australian continent even the desert areas will blossom and become productive and the country will sustain life forever. That is entirely in the hands of the Australian people and should decide by themselves to chose between life and permanent death of their people and continent. New Zealand will not be affected in anyway and will not change much for the next two thousand and five hundred years.

China will declare Christianity as a valid religion before the next ten years. In the next two years overwhelming parts of Chine will desire to know Jesus. Nothing will take place for a long time and suddenly many Chinese will hail Christ as their Lord and king and nearly all Chinese will become Christians. And all these will happen in not long distance. China will be a heaven in the entire earth and will deliver good things to the world. No decision shall be made to fight any of its neighbors and will desire to help the disadvantaged ones. Heaven will look like China. However, the entire nation will breakdown into ten distinct nations after five hundred years and China will lose its image and never will it rise again as a great nation. Knowing that the Chinese will desire to live in harmony with the rest of the world until two thousand and five hundred years. Then China will reunite and will become engrossed with hatred. Heavy war will rise and China and Europe will be involved in the fight. They will exterminate each other using deadly and lethal nuclear weapons and die because of the radiation that will disperse in the two nations.

Europe will be one nation in a maximum of thirty years including Russia. German will be their main language in two hundred years and will replace English. In five hundred years German will be a universal language and will dominate the entire earth. Even Chinese will learn German. Europe will prosper for the next two thousand years and will not show significant changes during that period of time. The environment will be dealt a blow and in the next two hundred years Europe will never see snow flakes but will still have plenty rainfall to sustain comfortable life in the continent.

Their population will constantly be replenished by migrants from other continents and the mixture of Europe will be different from its present state. However, the European continent will continue to be a viable and prosperous part of the earth. As time goes on mini languages will disappear in totality. Many great inventions will come out of Europe and great minds will rise from its population. Nothing bad will happen for the next two thousand years in Europe. Britain will become an integral part of Europe also and will never desire to make a separate decision different from its European counterparts.

America will never be a power for too long. Americans will desire to revive their old spirit but the damage is already done and will require tremendous work and prayer to restore their old spirit, patriotism and love of the nation. Americans have lost faith in their nation but they don't know it until the truth is reveals it to them.

However, America will not relent until it achieves its objectives and liberate its citizens from the mental slavery the Jews had inflicted upon them. As a result, America will become a very prosperous nation and a delightful country with God and Jesus at the center of their hearts. Faith will blossom in America and healers and great preachers will emerge from its population and their constitution will reflect that. A small part of America will become desert but will restore it by planting trees. Knowing that they will deliver messages of love and unity throughout the world and they shall be known as peacemakers.

Knowing that America will give, many, many, many help to Africa to save it from extinction. God's own nation will truly reflect the love of God and the American people shall be blessed for the next two thousand and five hundred years and will see absolute peace throughout that period of time. In the not too distant the borders between Canada and Mexico will become a thing of the past. America will become a much bigger nation. God will be their king and Jesus their prince. Nothing bad for the next two thousand and five hundred years.

South America will show dramatic change also. The deforestation of the Amazon will have a lasting impact in the continent. In two hundred years one-third of the continent will not sustain life. As a result, the entire population will be concentrated in other viable areas of the continent. South America will never become a single nation and will continue to maintain their separateness for the next two thousand five hundred years. Any attempt to unite will end up in total failure.

Some of the nations will fight among themselves but will be dealt heavy blow by the rest of the world. Soon, any nation that will start a war will be confronted by the entire world and squashed quickly. No war for the next two thousand and five hundred years. South American population will decline very little and will continue to be a fairly populated continent. Never will it become a power. Hatred will never prevail among the different nations and good cooperation will blossom but will never become one nation like Africa and Europe. Many bad things will not happen only massive earthquakes that will ravage mainly Chile and destroy many lives. Central America will continue to exist as separate nations. Honduras only will be flooded by the sea and lose a chunk of its land. Others will exist as they are today with little transformation.

The Caribbean islands will remain intact and no visible changes will take place in them. One percent of the Pacific islands will sink into the ocean. Hawaii will not be affected at all. Madagascar will become a very prosperous nation and rise to be a great nation mainly by tourism. Vietnam and other islands in Asia, with no Philippines, will become great allies with China and will join in the war against Europe and will be destroyed permanently. However, nothing will happen to them for the next two thousand five hundred years. Philippines will not be densely populated after two hundred years. There will be five different tsunamis that will wipe out entire islands. India will become a very prosperous nation in the next two hundred years. However, it shall break up into two nations barely two hundred years from now. Meanwhile two nations will reunite with India. Bangladesh and Sri Lanka will join India and become part of it in the next two hundred years. Pakistan and Afghanistan will unite and become one nation in no less than two hundred years. Every other nation in the world will not show noticeable change worth mentioning by the writer.

Environmentalists have no clue about the transformation of the ozone layer. Many assume that the discharges from cars and smokestacks and coal burning are responsible for the depletion of the ozone layer. That can't be much further from the truth. Their assumptions are totally untrue. First and foremost discharges from cars and smokestacks are too heavy and can not rise above a maximum of two kilometers into the sky and could never reach the ozone layer. The ozone layer is made up of the ozone air emitted exclusively by trees.

Trees absorb oxygen and emit carbon dioxide and ozone gas as byproduct. The ozone gas is extremely light and ascends to the sky at a very low speed and eventually ends up at the ozone layer. It takes it up to six days to ascend to such a height. As it ascends higher and high it slows greatly and when it gets to the ozone layer it is almost at a very low and stagnant speed. Even then, it doesn't stop permanently. It continues to move at an extremely low speed and eventually disperses into the universe. It takes almost two to six months for the ozone gas to disperse outside the earth's sphere. It rises almost two to three hundred miles above the earth surface during that time.

When trees emit ozone gas as a byproduct, the gas rises above the surface of the earth and ascends to the sky gradually. On its way it intercepts clouds which are composed basically of moist air. The ozone gas has the tendency to collide with moist air. Incessant collisions result in clusters and form tiny droplets. The droplets grow larger and larger and become heavier. Eventually numerous clusters of ozone gases and moist air form large and distinct droplets that fall as rain on the earth. When the droplets touch the surface of the earth they disintegrate and the moisture is absorbed by the ground but the ozone gases are released again and begin to ascend to the sky again until they reach the ozone layer and after a while dissipate into the universe.

Every tree of whatever type produces ozone gas a byproduct. Plants and grass absorb oxygen and emit only carbon dioxide and don't produce ozone gas. Only trees do. The more the number of branches and foliages the more ozone gas a tree produces. If one builds a cabin above trees he develops dizziness and becomes disoriented after sometime and could lose conscience. Ozone gas is deadly lethal. Astronauts cannot last long if they remove their oxygen masks and inhale the high concentration of ozone gas in the outer space of the earth atmosphere.

Humans don't come in contact with ozone gas emitted from trees because it never descends to the surface of the earth and always rises and travels higher when released from the trees. Never ever. It is a very light gas and flows higher and higher until it reaches the ozone layer. Forested lands produce tremendous amount of ozone gas and the concentration of ozone gas could be as low as two kilometers from the ground and to as high as three hundred kilometers. No moist air or cloud could pass such a concentration of ozone gas. It always ends in a rainfall and heavy downpour. That is the reason why there is always heavy rainfall in highly forested regions of the world, such as the Amazon area.

The reason why rains don't fall in the Sahara desert is because there is no concentration of ozone gas emitted by trees and as a result, the clouds always pass above the desert without any downpour. That explains the reason why thick clouds pass through the desert without producing a single drop of rain. However, thick forestation has its drawbacks. Very high concentration of ozone gas emitted by regions such as the Amazon forest deflect rich burning rays from the sun that are very valuable to proper functioning of the body.

Humans living in vastly forested area are deprived of burning rays from the sun have less dense bones, badly damaged organs and look deformed and don't attain the stature of a normal human being. Tribes in the Amazon forest are clear examples. People living in the Sahara desert receive constant burning rays of the sun are usually very tall, healthy and no defective internal organs. Little people like the Chinese and Japanese are deprived of the sun's burning rays in their history and no other reason. Africans living in the rain forests are also short and governed by short life. Asians living in rain forests don't live long but those living in other areas enjoy longevity because of their good diets which constantly replenish their bodies. If a person living in the Sahara desert is fed Japanese diet he would live to be more than a hundred years.

Depletion of the ozone layer is exclusively caused by cutting trees. As forests disappear so also goes the emission of ozone gas with it. There is no other explanation. It is a man made disaster and could be resolved and reversed very easily. Fumes from cars and smokestacks come down as acid rains and are toxic to breath but have no relationship to the depletion of the ozone layer whatsoever. No one knows of the ozone gas until now and nearly all experiments made their conclusions by observing plants and grass which don't emit ozone gas.

Ozone gases travel from lowlands to highlands before ascending to the sky. Their ways are opposite to the flow of water. Water flows from high to low altitudes but ozone gas flows from low to high altitude. Hence, if there is a forested area in a lowland the ozone gas travels to the higher grounds before ascending higher and remains concentrated above the highland. That is why rains fall in higher lands even if the forests are in low lands of the same area. Ozone gas hates dump air and prefers colder air to ascend to the ozone layer. Dump air sticks to the ozone gas and becomes dew after they form clusters. Every ozone gas desires to return to the ozone layer and avoids being trapped by moist air. Dirty air does not prevent ozone gas from ascending to the ozone layer.

Many reasons are given for the depletion of the ozone layer and none of them are true. Really no one knows when there is a depletion mainly because there is no any stationary concentration of ozone gas in the entire sphere of the earth. Ozone gas really never stays entirely in the earths atmosphere. More and more ozone gas are produced on earth and more and more of the gas escape to the outer universe. It is a constant cycle and never was a consistent concentration of the gas around the earth's atmosphere.

Knowing what we know we should plant more trees to produce more ozone gas and let more rain fall in larger areas of the world. Daily production of ozone gas is dwindling and that should be a concern to the world. Very soon, many parts of the world will live in dry areas and desert their local habitations as a result. No one will devise any method to reverse the depletion of ozone gas. The only solution is to plant more and more trees. Every nation should invest heavily in tree planting programs. If no one does that the whole world will become desert in no time and the human specie and animals will perish.

The ozone gas molecules act like glue and attach upon them moist air molecules. One ozone gas molecule could have attached to it up to a maximum of two hundred moist air molecules. The newly formed compound collides with another similar compound and stick to one another. And the process goes on and on until a droplet is formed. The droplet grows bigger and bigger until it is heavy and falls to the earth as a raindrop. The ozone gas is just a catalyst and is much tinier than the moist air molecule and has no role in the composition of the rain.

When the droplets fall to the ground the ozone gas is liberated when the water is absorbed by the ground and beings its ascent to the sky once again. That is why fresh rain has a smell that lasts for a short period of time. Never was a time when ozone gas remained on the earth's surface for a long time. It always made its way to the sky as soon as it could. Humans never noticed of its existence because of its short duration of stay on the surface of the earth. Ozone gas could stay up to six months on the outer sphere of the earth atmosphere before it dissipates to the universe. That is why constant new supply of ozone gas is essential to sustain life on earth.

Never will there be too much ozone gas because it regulates itself and those areas that produce much concentration of ozone gas have a much higher and faster rate of dissipation of the gas into the universe. Only that those who live in rainforests have always a good concentration of ozone gas in their atmosphere and they deflect valuable burning rays of the sun that are relevant to their existence. Other than that there is no impact on the vegetation for plants and trees don't require the burning rays of the sun to survive. Only living beings. Man-made damages have depleted the production and level of ozone on the earth atmosphere.

Those that have no production of ozone gas, mainly the deserts, occasionally have rainfalls because the ozone gas migrates very slowly and accumulates there and takes it far too long to disappear into the universe. When clouds pass by they are converted to rain. Then there would be no rain again until small ozone gases accumulate and are concentrated enough to force moist air to be formed into rain. Never was there consecutive days of rainfall in the desert. Never will there be any relief until the entire earth atmosphere relies on good concentration of ozone gas. Mankind's desire should be to fill every inch of the surface of the earth with ozone gas. That way, every part of the earth would be productive and arid lands will disappear.

Healing the earth should be the priority of every living being and deserts should be considered as affronts to humanity. There are many determined nations prepared to fight the constant depletion of the ozone layer. All these nations have to do is plant more and more trees in different part of the world and the problem is solved permanently and humanity will thrive and prosper. Many nations will believe in this message and act immediately and will have nothing to regret about. All those that would deal with ozone depletion will become a blessing to their neighboring nations. Ozone depletion is not one nation's problem. It requires collective effort to deal with it. Every day that is wasted is a death sentence to mother earth.

Good nations will become great motivators to other nations and those that betray mother earth will be made to look like evil nations. God desired to reveal and heal the earth because of the love of his Son to Humanity. Jesus died for the sake and to save humans from extinction and his love for the world is unprecedented. Nations will thrive and prosper if they really decipher the messages of Jesus and anoint him as their Lord and king. His name is power and is the only ticket to the heart of God. Those who don't believe in Jesus are just base humans and delve into the rejection of God. Grave consequence will be felt from now on those nations that desire nothing to do with the heavenly Son of God. The law of God requires that you delight in his Son and know him only through his beloved almighty.

Never was any man relegated to hell for knowing and worshipping the Son of God. Many fail but a few succeed. Healing the mind is a process and begins with knowing Jesus as a redeemer and savior. Man doesn't know that evilness is an instinct embedded in his heart. No one wipes out that evilness from his heart apart from the blessings of Jesus Christ. God knows no one that does not come to him through the name of his glorious Son. Man desires to communicate with God neglecting the name of Jesus Christ. That is a vain desire and never happened. Those that revere the Lord and have no desire to approach through his Son are merely wasting their precious time. They might as well not pray. However, God does not relegate them to hell but can only go as far as purgatory. Given that men and nations should scramble to heal and get maximum blessings and earn everlasting prosperity by merely acknowledging Jesus as their daily bread and butter. No one makes an error by doing that and they will reap abundantly for their loyalty to Jesus. God hears them and fulfils the desires of their hearts. No one who believes in Jesus despises his fellow man and no one inflicts pain, torture, torment or death upon others. All those who do so knowingly find hell fire at their expiration.

Give and you shall be given, grant and you shall be granted, maim and you shall be maimed. Every man makes no effort to walk strictly in the days and nights of his life and makes his decisions on the dictates of his mind. That is a God given right bestowed upon Adam at the beginning of creation. Man has absolute right to decide his own destiny. God does not interfere. Those who decide to follow God from early childhood delight the heart of God and are compensated and become the sons and daughters of the Lord. Hell is never an option to them. Many will die and become archangels and will delight in the presence of God to eternity. Not many improve to that level. However, those who deliberately reject God and live a void life are immersed in the dictates of Satan and his evil messengers. None of them is healed and they burn in hell fire at their death.

God's laws are sacred and never, never change. Those who break the laws of God are in danger of the wrath of God. Great things take place when nations hail God and do things in the manners of his laws. Every one who desires and makes much effort to know God and his Son, Jesus Christ, have the Holy Spirit in them. Those who know God are never independent and are constantly under the bondage of the Lord. Any deviation they are convicted by the Holy Spirit and know it in their hearts. Man has the choice to enter into the domain of God's kingdom or reject it in its entirety. Either live in absolute freedom and liberty in the domains of Satan or become the subject of God and be dictated to by the Holy Spirit.

Nothing is directed by God. Man makes his final decision and the direction he would follow and that is his inalienable right that was granted by God. God did not strike Adam with lightning to prevent him from partaking in eating from the forbidden fruit. God gave his instructions but it was entirely under Adams discretion whether to obey or reject the dictates of God. However those who repeat the same prayer repeatedly over and over they deafen the ears of God and the Lord clogs his ears from their hearings.

The problem arises when he dies. Those who lived under the constant bondage of God because of their faith in God and his blessed Son, Jesus Christ, live in absolute liberty and freedom in paradise and to eternity and become angels. Jupiter from earth is just a walking distance for an angel and could be compared to a casual walk from ones home to a nearby convenience store. No more messages of the Holy Spirit and nothing to impede any wishes of an angel. Giving that angels deliver no sickly messages in their minds and live in the heavenly realms of God's kingdom. Messages in their heads are known to God and messages from God are delivered in their minds. Minds and souls depart from a dead person and upon entry into paradise acquire angelic forms in a millionth of a second. When the sperm of a male meets the egg of a woman and collide a soul that has a mind is created instantaneously and in a millionth of a second. That is when life begins.

It takes nine month for the body to form and become a baby and then develop into a man or woman. That same soul with its mind departs instantaneously and acquires angelic form. It takes nine months to form into a complete human but takes a millionth of a second for that same soul and mind to transform into a complete angel. Having said that all angels are different from each other just as humans are. No identical angels ever. All angels know all their activities on earth. Paradise is an extension of earth and death is only a temporary setback that lasts for a fraction of a second.

When the soul departs from the body upon death it is filled with a wealth of information. The mind of the soul knows exactly all its activities in its entire existence, from the time the male sperm met the female egg and unto death. When it departs from the body no one would tells the soul where to go. An evil soul knows precisely where to go and is immersed in hell while the good soul enters paradise. In other words the soul judges itself. Only the appropriate door is open to the soul and in accordance with its activities on earth. To an evil soul only the gates of hell are open and to the faithful soul only the gates of Paradise are open. The soul has the knowledge of every millionth of a second its activities while on earth.

Having said that not all angels remain loyal to the Lord in paradise. Some of them desire to return to earth for they miss their lives on earth. Those angels quickly find themselves in perdition. Perdition is a secluded periphery of hell and is guarded by loyal angels. It is like prison and is a temporary detention center. Angels regret and confess and plead for mercy and after a while, are returned to heaven. While in perdition they have a good glance of hell and the torment of those in it. As a result, no angel that went out and returned ever desired to depart from paradise.

Everyone in paradise is knowledgeable of their past life while on earth. No one knows any members of their families even if they are in paradise with them. That knowledge is never desired by angels and don't need family members. Angels live in collaboration with each other and dare not give any hateful feelings in their minds. So many of them are constantly thrown out of heaven after they desire evil things in their minds. Not every angel is thrown to hell but some do.

Overwhelming majority of them are sent to purgatory and from there make their way back to paradise. Purgatory is a place of constant repentance, regrets and great prayers. Those who inhabit purgatory are in constant turmoil and quarrel with one another in a bid to gain the favors of God. Those who show enough regret and repent with the sincerity of their minds are promoted to consolation after serving for sometime. Consolation is an orientation place and just a step before paradise. Angels administer to those in consolation constantly and rehearse to them their rights and obligations when they enter paradise. Those souls that find themselves in purgatory or consolation have heavenly bodies, different from that of angels or Satan.

Every time an angel desires a name the man or woman appears in front of him. Angels have constant knowledge of humans on earth. Every human being is monitored by his own angel. Those who follow in the path of God the angels dictate the aspiration of God in their hearts. Those who reject God remain out of the domain of the angels and are monitored by the devils instead. However, many go from doing good to doing evil and vice versa constantly. When they do that the devil desires to maim them and inflict deadly things in their lives. Many become the devil's messengers and then desire to hurt others and inflict pain upon them eventually.

The angels become helpless and desperately desire to recover them but those who stick to the devil become severely evil people and deadly human beings. The devil is known to have many powers and desires nothing good for humanity. Heavenly powers are known to ache when one person rejects God and becomes the subject of the devil. Many great minds have greatly erred and deserted God and believed in evil things and no one desired to live in the ways of God. Those who do that discard God and formulate their own theories beginning from the creation of mankind to their destinies after they die. None of them knew the truth and none of them made it to purgatory, consolation or paradise. Hell fire is to those who formulate unfounded messages to discredit the existence of God.

Many of those who discard God entirely and live in total freedom and liberty do not know that they do so at grave danger to their eternal existence. Satan messages his secret service agents, the demons, to persuade a man to reject God and enter the domain of Satan. The deserter loses hall heavenly blessings and lives as a subject of Satan. And Satan never bothers him again and the unbeliever lives according to his own wishes and desires with no interference from Satan while living on earth. His heart is totally freed from the dictates of God and his commandments and lives in absolute freedom and liberty. While believers are constantly monitored, convicted by the Holy Spirit in the case of Christians and filled with daily messages in their hearts and souls, the unbeliever lives devoid of all those burdens.

Everything his heart is filled with is what he believes in. Great world thinkers are devoid of God's inspirations and formulate theories according to their thinking. Saying that, overwhelming number of them don't decipher the real truth and deliver messages that are purely fantasies and bogus. Their brains become very creative and their minds delve in unfounded messages that are purely of their own making and no guidance from anywhere. They get no messages from their angels and Satan never consults with them knowing that they are his subjects and will enter nowhere but his domain when they die. Knowing that, many unbelievers believe they have a duty to mankind and do everything they can to deliver messages to mankind. Unfortunately, all their messages are deadly threats to the ultimate destinies of those who follow their thinking.

Many of the messages are fascinating theories and mind bugling and many followers live and are guided by such beliefs all their lives. Many unbelievers who reject God are "free thinkers" "free minded" "free spirited" and have great worldly minds compared to believers and believe deep in their hearts that they own the world, and that mankind should be guided by their own philosophies. And many of them have succeeded. A lot of them shred into pieces the perception of God and convincingly argue the invalidity of the existence of God and establish great followers as a result. They are lawyers, teachers, husbands and wives, medical professionals, engineers, ordinary citizens etc.

Great minds are nothing but the very daring among us who discard the message of God and delight in the sensations of the messages deciphered by their brains. Nothing special. That is all they are. That is nice and dandy and after all man is bestowed with the ultimate authority to decide his own destiny by the voice of God. Knowing that, nothing formulated by such believers is designed for the destruction of mankind and many of them formulate theories and methods of how to free oneself from the bondage of God and live life in the world in the best way possible. That is why many followers desire everything but God.

Many followers of such people are happy and live in the happiness and comfort and would desire nothing else. And there are a lot of them on this earth. There is probably no better way to ascertain ones freedom and liberty better than discarding God, become the subject of Satan and experience no interference in your life. That is indeed absolute freedom in the real sense of it. Having said that, everyone that has no messages of God in his mind remains the same for life and would not exchange his liberty for anything. That is one way man makes his decision and has to live with that decision to eternity.

Those who have angels become subservient to the dictates of God and know that their lives are governed by his messages. To the unbeliever that is insane and are delighted that they are not like them. Messages from angels are made known to the minds of believers that influence their way of life and delete any other perceptions that creep in their in their minds. Greatly religious people desire nothing on earth and look forward to their lives in life thereafter. The unbeliever does not mention life after death and many of them believe that when they die it is the end of them.

Many unbelievers speak of the abundance of life on earth and nearly all the time delete any perceptions of heavenly existence. Knowing that, heaven never closes the door on them. Those who realize the truth of the existence of God are never rejected and God never desires to destroy any being who knows his name. Man is made in the image of God and the Lord aches whenever one man strays from knowing and worshipping him. Man was created to desire God and praise his name. That was all he was created for and nothing more and nothing less.

When man believes in the none existence of God and desires to do nothing with him, Satan delights and knows his domain is growing. Satan was not worshipping God and that is why he was driven out of the heavens. Satan desired to do nothing with God when he was an angel in paradise. He was the first angel to rebel against the Lord and never was another angel that held such a messages in their minds. There was no hell in the heavens prior to the rebellion of Satan. Hell was created by God to house Satan and to torment him for his evilness. Hence, Satan is just a rebellious angel who did not believe in praising the Lord. When you don't worship the Lord you delight in the deeds of Satan and nearly all the time end up as his subject. Hell was not a terrible place at the beginning. It was a dark, gloomy and dreary place but Satan was not engulfed by fire. Just a prisoner.

Hatred not to worship God became the trait of Satan and those who do so are nothing but the messengers of Satan. Satan never lost his angelic powers. He is still as powerful as the angels of the Lord in paradise. Angels have the power to communicate with two hundred thousand members of paradise simultaneously and so does Satan. Angels are never lonely and are in constant contact with each and spend their days talking, joking and exchanging ideas while doing their assigned duties. Satan can, similarly, have presence in the minds of two hundred thousand people at a single time.

At the beginning, when Adam was created very one that died entered into heaven regardless of their activities on earth. Satan was the only lone tenant of hell. As time went on some of the angels began to exhibit sickly activities in heaven in continuation of their lives when they were on earth. The Lord was amazed. How could they live their earthly lives when they entered heaven? As a result, the Lord decreed that none of those who desired evilness in their hearts should be allowed to enter heaven again. He saw Satan sitting lonely and hence made it a law that those discarded from entering heaven would become a company to Satan and remain his subjects to eternity.

And things went well for many millenniums and the Lord was glad. However, some angels committed minor offences in heaven repeating the things they excelled in while on earth and that was causing distress in the minds of God. Hence, the Lord gave the power to the angels to arraign such deviant angels, try them and convict them. Those found guilty were relegated to hell. The convicted angels lost all their power of an angel and wore the body of heavenly bodies. They have the power of only a single soul that would depart from earth upon death. These petty criminals became to be known as demons.

They were mostly liars, crooks, deceptionists and meddled in other angels affairs. A demon could only affect the mind of a single human being on earth but sometimes be in clusters when they invade a favorable candidate. Demons perform tricks and deceptions on those they possess making a person delusional and turn his head upside down. When a person is close to rejecting God they inform Satan and Satan does the final finishing touches to influence the minds of such people and makes them his subjects permanently. Demons are loyal to Satan and they revere him as their master and lord. No demon desires to receive any messages from the Lord and neither do they get any.

After a long time, the angels discarded the petty criminals regularly and heaven was in peace and there was perfect harmony among the angels. Presently, there are over two hundred thousand demons that infest the earth since the creation of man. Sadly, most of them are in Africa and Asia where their tricks and deceptions yield much fruit and they recruit many subjects for Satan. Having said that, no demon desires to know much about the person they invade and place themselves near his heart. Demons get excited when one man does evil things. That is a message for invasion. The man is invaded with messages to influence him to commit more tragic things. Heavily demon dominated individuals become extremely and sickly wicked and evil. Demons have no mercy for individuals they invade and possess and ultimately lead them to their final destructions. Many demons stay close to an individual for very long time waiting for him to make a grave error and then pounce in immediately.

They know that grave errors are a sign of the person' drift towards the domain of Satan. Demons never ever give messages to the individuals they possess. They merely make damaging whispers in the periphery of the hearts of the person. No demon has the capacity to enter a man's heart. Demons know when to stay and when to leave. They try their utmost best to influence the heart of a person and frustrated and when they don't succeed they curse that person and depart and no demon tries to influence that person again unless that person commits very, very terrible thing. Demons don't have the capacity to know what the demon next to him is thinking. As a result, they operate independent of each other and only do similar things when they are in cluster. Heavy messages are not imparted by demons. Only deadly, sickly, demented and derailed whispers are emitted from them and those who adhere to them portray such traits. Demons are mere nuisances and wield no special powers.

One day, the Lord noted a very evil angel that contemplated and desired to harm other angels and that angel instantly popped in hell. When the wicked angel realized he was in hell, and knew the reason for his banishment, he began to weep and pleaded sincerely for pardon and mercy from God. The Lord noted his appeal. God restored one-third of his angelic power and instead of having heavenly body he let him have features of man and animal for he was wicked.

He had goat's horns on his head, had hands of man, legs of midget, feet of donkey hoofs, belly of snake and other gruesome features. And that became the devil. All the angels knew what had happened to the wicked angel amongst them and no angel plotted such wickedness again. Hence, the devil is only one but with a third of angelic powers. Devil never recognized Satan as his master and never worshipped him. He has an independent mind and dearly loves God for he never plotted against God while in heaven. Even demons and Satan revere God. If they don't, they would have been exterminated a long time ago.

Devil is hateful spirit with despicable features. Those that had a glance of the devil remained traumatized for life. No one want to deal with the devil for his acts are sickly evil and wicked. The devil enters a home and never leaves until he devastates entire generations. Heavenly wickedness is in his head and nothing as terrible as the devil. Devil maims and tortures and desires nothing good for mankind. Devil worshippers deliver mostly animal blood sacrifices to appease him. But the devil goes further and demands human sacrifices to reward his worshippers.

If a witchdoctor doesn't provide human sacrifice the devil turns against him and sucks his blood instead. The devil delights in torture and torment. Heavenly bodies can not be compared with his deadly, sickly hateful and diabolic messages. The devil's messages are stored in the minds of the people he cripples and never go away. Once a devil enter you, you become a devil and repeat his deeds for life. He searches for clients knowing they don't like the Lord. The devil knows the minds of the people he inflicts pains on just like an angel has full knowledge of the thoughts of man. The devil has a third of angelic power and could inflict his torment on tens of thousand of people simultaneously at any given time. Africa is the base of the devil followed by India in the world. Voodoo is nothing but devil worshipping. All the religions in the world except Islam and Christianity are all devil worshippers. They may vary in the methods of sacrifices but all of them do so to appease the devil. Moslems and some Christians make animal sacrifices but that is a tradition of Abraham as dictated by God and not the devil.

Many times Satan desires to consult with the devil and not to harm his subjects on earth who had already rejected God and were living as independent and liberated lives. The devil, however, always rejected the requests of Satan and warned him not to interfere with his worshippers. As a result, there is always a clash between Satan and the devil over an unbeliever who had rejected God. Satan guarded his own and the devil guarded his own. An unbeliever is torn between the evil deeds of the devil and the sickly messages of Satan. Many unbelievers dish out the evil deeds of the devil and profess disgusting messages of Satan. Such people are good for nothing and are always despondent and don't understand why their minds go back and forth.

When hell was formed and Satan was relegated into it the Lord knew that he did a good thing. Then all those who made grievous errors on earth were directed to go to hell to be in Satan's company. Soon, hell was filling up and not many people entered heaven. The Lord knew why hell was filling up and made his decree. From then onward all those souls that believed in God while on earth but made major errors were to go to purgatory. Hence, purgatory was created, very far away from hell and paradise. Those who cleansed themselves from the errors they committed while on earth and entered purgatory were delivered to paradise immediately they showed enough regrets and sincere repentance. And the Lord was pleased with his decision for he knew that it was not fair for the soul that believed in his name to go to hell.

However, some of those souls that were elevated from purgatory to paradise and became angels were not too disciplined and rained chaos in paradise. While in purgatory, the souls constantly fought and beat on each other and dealt big blows to their cellmates. When they entered paradise, the never shed their ways in purgatory and behaved abnormally in paradise. As a result, the Lord decreed again and consolation was created. Consolation became a transit station where angels from paradise lectured the heavenly bodies that inhabited it and prepared them for their eternal existence in paradise.

There, the souls remembered their state of condition in purgatory and never resorted into fracas. They stayed in consolation until the angels decide that they have been reformed enough and suitable to exist in paradise in harmony with the other angels. Once found clean they were promoted to paradise and attained the form of angels. Consolation also became the destination of faithful people who committed minor error while they existed on earth. Such souls skipped purgatory and went directly to consolation, just a step away from paradise.

When the body dies and the soul departs and enters paradise it remained in the periphery of paradise and doesn't make it to the real paradise. The next step is to be brought before Jesus Christ to determine its final eligibility to enter paradise. The angel of the soul that had been monitoring the person's actives while he lived on earth read his life history to Jesus. Then Jesus made the final judgment. Those accepted by Jesus to be righteous with no blemish are immediately ushered through the gate to paradise. Those that Jesus finds not cleansed enough are sent to purgatory for further reform, repentance and regret of their activities on earth. Jesus never desires to send anyone to purgatory but the angels deliver sickly messages that were committed by the soul while existing on earth. Jesus sends them with heavy heart and complements them for being righteous and delivers good messages unto them. He tells them that it is the law of God that he is enforcing and that he delights in them and wishes they would return to him soonest after their period in purgatory.

Heavy messages are not in the minds of those in purgatory. The souls understand precisely their sins and the reasons why they found themselves in purgatory. They shed their sins one by one when they would be deleted by the Lord and they depart from their hearts. When a sin is forgiven by God, the soul doesn't delve on it anymore. It goes to the next sin and begs for forgiveness from the Lord. The souls knows precisely when his plea is heard and his sin is forgiven by the Lord. When they realize that they have one or two sins left to be forgiven they begin to prepare to enter into consolation. They go through the same process until they are brought to the presence of Jesus eventually. Again, Jesus decides their final fate. If they are still not fully cleansed Jesus sends them to purgatory but mostly for a short period of time. One has to be absolutely clean to enter paradise.

When God initially created hell and sent the disobedient souls to inhabit it he was pleased with what Satan was doing. Satan organized hell and dealt no evil messages to them. He felt fulfilled and maimed no one and tortured no one. The God told Satan that from then onward hell would be known as Satan's domain and put all its inhabitants under his domain. Even more, God granted Satan the ability to chose those rejects and dejects who knew him not while they lived on earth. Satan told the Lord that he would make the whole world follow him with such a power.

The Lord dared him to do so for God knew the lost and the found and not many found God unacceptable. Satan was heavily desiring to prove God wrong and hence, became the power that he is today and began to influence numerous and lead them to reject God and become his subjects. However, when Satan found that there were not many following him he searched the mind of God and the Lord appeared to Satan. Satan told God that the hearts of men were not following him and hence, asked God to let him govern the earth and discipline men in his own way. God saw that Satan was doing a good job in hell and the Lord granted him his desire. Hence, Satan became the prince of the world.

Many millenniums passed and the Lord saw that not many people on earth were desiring him any more. He glanced over the earth and saw sickly things happening on the earth. Men worshipped boulders and trees and hated the name of God. The Lord knew that it was the handwork of Satan. The Lord inquired from Satan and Satan told him that he was the prince of the earth and he desired the people to worship him. God knew that he made an error but wouldn't turn his word. From the onward God knew that Satan never desired to do good things anymore but rather became an evil one. God wished to destroy Satan but he thought against it for Satan never denied the Lord as his God.

Satan dealt a big blow on God's reputation and delegated authority all over the world replacing God. His head was swelling and began to see himself as junior God. Having said that Satan never wavered from his stance never to worship God, the main reason why he was thrown away from heaven. Immediately, God made a decision to prevent Satan from entering the hearts of men. So, Satan never enters the heart of mankind. They only way he could influence man is by accessing only the periphery of the heart. So Satan lost popularity because he was stripped of his power and was no more residing in the hearts of men.

Satan didn't know what to do for he was losing many to purgatory, consolation and paradise. He made a very costly error when he told God that people were discarding God and worshipping him instead. He waited for a very long time and then realized that many were coming to him once again. Then he noted that those who did so were disappointed with God. He saw that the entire world was wailing and yelling for a different God. He wanted to know how it happened so suddenly. Then he knew the reason. He found out that no one wanted anything to do with God and had discovered that Satan was a very delightful angel and were exclusively worshipping him. He knew how it happened. He never thought it would have so much effect but he went around the world revealing himself everywhere he went. That was the reason he figured out. No one saw God but they could see him.

God saw what Satan had done and was enraged. He torched Satan's face with fire and he was engulfed in fire. Satan wailed and cried in pain and agony. He finally killed the fire but he was defaced beyond recognition. Satan had an angelic body but his face was so gruesome and terribly scary. No one was ever as ugly as Satan. Knowing his appearance Satan never went around the world and revealed himself to mankind again. Those who saw Satan died with fear. God never felt any regret for destroying Satan's face.

Many millenniums later very powerful message came to Satan. The entire angels wanted the Lord to destroy Satan but the Lord was reluctant to do it. He got some relief and was very cautious from then onward not to offend God. Very rarely did Satan made decisions that were not endorsed by God. Immediately after he had a defaced appearance Satan made a decision to revenge on his subjects. He went to hell and what he saw enraged him. Those in hell were delighted and were involved in conversations, laughed and enjoyed their life.

Satan was never happy since his face was marred. He couldn't stand those who were his subjects ridiculing his appearance. He immediately torched everyone of them and saw them suffering in agony and his heart was delighted. The entire hell was filled with wailing subjects of Satan. He killed the flames but left the embers to torment them. Everyone of them saw that Satan was desiring to exterminate them hence, they began to plead for the help of God. God heard their wailings and interviewed Satan. Satan told God that they were his subjects and that was delegated upon him by the word of God. He told God that he had the authority over everything but hell. As a result, he had the prerogative to determine how his domain should be run. God knew that he had made an error but could not take back his word. As a result, Satan began to torment his subjects and the Lord was helpless to change that for his word is final.

Satan saw that the torched heavily bodies in hell were no longer wailing but just groaning and sighing and breathing heavily in great agony and extreme pain. Then he desired them to worship him while they burned. They reluctantly whispered praised of Satan in the midst of their torment and that angered Satan. Hence, he desired to torment them even more. He pricked the hearts of those who did not constantly worship him in a loud voice. That is what Hell looks like until now. Those who go to hell find themselves torched the second they enter its gates and would worship Satan for every second of their lives to eternity.

The Lord made daily visits to the world for a day is a thousand years to the Lord. One day, God waited and waited for those who believed in him to respond when he soared over the earth but none of them reacted. Rarely did anyone flinch. The Lord knew what to do. He showered them with filthy waters and the rain went on for twenty days. Many drowned, however, the people set their hearts to seek the Lord. Never would I do that again said the Lord. He regretted his word afterwards when he saw them return to their old ways not many days later. So the Lord withdrew his statement. After many years passed God found that the world was corrupt and beyond redemption. He chose the only faithful single family that existed among humanity and ordered them to build an ark for he was determined to destroy the world with flood. The man of the chosen family was Noah.

God ordered Noah to build an ark and the Lord dictated its dimensions. So Noah with his two children and one grandson set to work on the ark. He bulldozed with his axe logs of wood and fastened them one to another tightly. The Lord showed him a huge flat piece of land adjacent to a very tall mountain. He tied logs to logs and covered the whole flat land in thirty-eight years. After that he covered the vast log filled land with grass. Then he built a little cabin in the center of the logs. Many saw him build the ark and were sickly disgusted. He told them the truth but they all scorned him and laughed at him. They said he was making up the story even if earlier many of them had drowned.

When the ark was completed heavenly deeds began to happen. Without Noah's knowledge the whole log field was infested with all sorts of animals. Noah wondered why all the animals were making their way towards the ark. Then he realized that it was to feed on the grass he placed upon the logs. Those who fed on grass were to come first and then those who fed on the animals followed them for the field was extremely large. He saw that virtually all the living animals that existed then were on the ark. Immediately the Lord poured heavy rain that went on for forty years. All the animals endured the rain and fed on the grass and those that devoured animals fed on them also. After the forty days the rain suddenly stopped and Noah went out to see the extent of the damage done by the heavy downpour and was alarmed. He could see no land as far as his eyes could see. Even the tall mountain had disappeared. He waited for two days and found no land. Another three days he saw a dove with a leaf on it mouth and he knew that the rain had dissipated and somewhere there was land.

The inhabitants of the town where Noah resided in had no knowledge that logs floated on water for there were no waters around them. If they had known they would have found safety on the ark. Noah didn't know why he was cutting logs and tying them one to another and only obeyed to the dictates of God. He never ever thought that logs floated on water and was surprised when him, his family and all the animals did not sink. The logs held firm and did not disintegrate. The whole field was one floating ship and his cabin the only protruding object above the water.

Many of the animals were barely visible after thirty days of rain for there were too many of them. Even Noah's cabin was flooded with water knee high. Many of the animals found grass floating on the water and others devoured those closest to them. In the end all the animals survived even if they were severely famished. Noah stored too much food that would last him and his family for only ten days. After that, he had to resort to slaughtering animals to survive and they fed on raw meat until the land dried. After five days, the water suddenly disappeared and the logs were on the flat land. Noah waited for another two days and then went to see the land and never saw a single surviving soul. They were the only living beings.

Many centuries passed and the descendants of Noah multiplied greatly. The Lord hailed them. After the next two centuries every one of them made a pledge to honor the Lord throughout their lives and they became righteous people. Every one desired to know the Lord in his heart and the Lord was pleased with them. The Lord had vowed not to destroy humanity by flood again to Noah and he kept his word until now. The people made many innovations for the Lord had opened their eyes. Every one of them had many skills and invented numerous things and their life was prosperous. Many tall high-rise buildings were erected and built many tall structures. Heavy pillars were raised with pulley systems and decorated them with terraces around them. God made them build magnificent mansions and dwelling places and lived in opulence. Grave sicknesses went away and made medicines from various kinds of herbs, leaves and roots. No one knew evilness and the Lord was revered in their hearts.

After two hundred years a sage member of the city of Babel proclaimed a vision to the people. He said that he saw a towering tower on top of the tallest flat surfaced mountain that was in the vicinity of the city. He told them that the tower touched the gate of heaven. The people wondered how they could build such a tower and began to plan their strategies. Soon work began and every one went and contributed his share daily. Nothing substantial took place for thirty years of the construction of the tower. Every one knew of the value of the tower and every one was eager to see heaven. Men took turns and no one was exempt from the labor to construct the tower. The tower had a large base but as it rose higher and higher it became thinner and thinner. The way to the top was a spiral around the tower. The body of the tower was made up of well formed bricks and densely packed together.

As the tower grew taller and taller the workers began to complain that they were chocking and running out of breath. As the tower grew taller some of the workers collapsed and died and only soon only a few people could climb to the summit and returned alive. It was not very long before numerous people chocked and nearly died. At last only a handful people of the city went to the top and returned safely. The sage man was consulted and he told them that they had nearly found the gate of heaven. Construction continued for a very brief moment until no one could go a step further before he chocked and nearly died. The sage man told them that they had finally reached the gate of heaven. And the people were very delighted and commended themselves for their marvelous efforts.

The sage man told them that if anyone was prepared to go to heaven should climb to the summit of the Tower of Babel. Nearly two years went by and no one desired to go to heaven. At the end of that time the old sage man told the people that he was prepared to go to heaven and he went with an accomplice to reach the summit of the mountain. Barely two thousands meters from the base of the tower the old sage gave up the ghost and died.

The accomplice returned and reported what had transpired. He told the curious crowd that surrounded him that as both of them climbed towards the top of the tower the old sage began to breath barely. They went further and the old man began to see angels and became delusional. He sat him down and then laid him on the ground. The old sage hyperventilated and took a last gasp of air and died with his mouth wide open. His accomplice told them that he saw his soul depart from him as the old man opened his mouth for the last time on earth. He asked him where he left his dead body and told them that he threw it from the top of the tower and unto the bottom of the mountain. They were all amazed and delighted in knowing that the old sage had found the gate to heaven and was with the Lord.

As time went on many people who desired to go to heaven climbed to the summit of the tower and all found their way to heaven or so they thought. Some went barely two thousand meter before they gave up the ghost others went a little higher but hardly any of them successfully reached the summit before their soul departed from them and entered through the supposed gate of heaven. Nearly all the elderly people were wiped out as a result. Then began the younger members of Babel and there were nearly two hundred thousand people that climbed the tower and released their souls to enter heaven in fifty years. Nobody mentioned how they departed and only told those on the ground that the dead people's souls had entered heaven. Everyone of them suffocated and suffocated and suffered tremendously before expiring. Those on the ground never knew of such sufferings and the trauma experienced by those who volunteered to enter heaven before they died. They all assumed it was a blissful death. The dead bodies were always thrown to the base of the mountain.

Many dead bodies souls arrived in hell before their appointed time and Satan alarmed the Lord. God saw what had taken place and wondered at the innovation of his creation. He delighted in their intentions but knew that they had violated the law of God. No human is allowed to take his own life and those who did ended up in hell. When man was initially impregnated the soul, which is a breath of God, entered the combination of egg and sperm in the woman's womb. When the baby's head pops up at birth and takes its first breath an angel instantly resides in his heart. A baby is virtually a body with angelic heart. The angel remains inside the child's heart until a maximum of thirteen years. After the age of thirteen years the angel departs and the teenager is given man's right to ultimate freedom and liberty with no interference from either God or angels. From that time onward the adult child makes his own decisions and lives by the consequences of his aspirations. Not all children have angels in their hearts until the age of thirteen. In some, the angel departs from as early as five years old when the child matures too fast and begins to assert his independent rights.

While in the child's heart the angel assesses everything about him/her. From such tabulation the angel determines for how long the child should live before he dies. That power is exclusively given to angels to determine. They follow God's formula to do so. When the expiry date is determined it never changes ever and under any circumstance. God has the ability to access, in blink of an eye, the life records of all humanities ever created and those that will be created to eternity. However, the Lord deleted all such records and doesn't access them because he had given his word to mankind that he would not interfere in their decision making process and had given man the absolute right of freedom and liberty to decide his destiny.

A pedophile that rapes a child is in essence raping an angel and he commits an unforgivable sin. No amount of prayer and invocation of Jesus' name could save him from hell fire. A pedophile's deeds even enrage Satan when he hears the wailing of the angel inside the child. As a result, Satan has a special place for pedophiles and throws them in the deeper part of hell. There, the pedophile not only burns in ember fire but sits in an inferno in the deep part of hell. Demons and Satan do not entice a pedophile to commit such an atrocity. They are exclusively influenced and guided by the devil. What the devil couldn't do while in paradise and his intention to hurt angels he does it on the earth as a vengeance against his hatred for angels. The only solution to a pedophile's problem is to hang him. The devil never departs from a pedophile and he never stops raping children until his death.

The Lord was measuring the Tower of Babel out of curiosity. It was thirty thousand cubit feet wide and three thousand meters high. The Lord made up his mind to desire a solution to the Babel of Tower problem. He made up his mind to divide them into many parts of the world. So he divided them and made them to desire to migrate to other lands. But the people did not do so. How dare them disobey my orders, said the Lord. Then he knew why it was so. The people had desire to migrate to other lands but didn't have the means to get there. So the Lord made them have the knowledge to build many huge ships to navigate the seas and oceans. Even then they didn't travel far. So the Lord said, I will confound their languages so they wouldn't understand each other. As a result, only members of their families and close friends understood each other and there were suddenly many languages. And still no migration.

The Lord looked at them with curiosity and wondered at their innovation. Man had angelic powers to communicate by telepathy and the people were doing so effectively and understanding each other. The Lord said, no more telepathy for mankind and he striped man of such powers from that day onward. Only animals have the power of telepathy. And the people's state was total confusion for no one knew what the other one was saying. Immediately, all the inhabitants of Babel went their separate ways and populated the entire earth.

A set of people travelled to the city of Mesopotamia and there they established themselves. Another set went another way and settled in Asia. Others travelled by ship and crossed the oceans and made their settlements in South and North America. So far, no one went to Europe and Australia. Heavenly messages went into their heads and guided them all their ways to travel into these area. Every one of them worshipped the Lord and the Lord guided their ways. No one made idols nor worshipped anything but God. They were all righteous people.

Having said that, every one did not delight in their new environments. They daily wished they were in Babel and its surroundings. They were highly civilized people and they lived in great comfort and opulence while they were in Babel and now they had to start from scratch. Heavy sadness was always in their heart and discerned the messages in their hearts daily. After many years they began to accept their new fate and tried to make the best of it. They all knew that they were not going to return to Babel and the thought began to dwindle gradually. Heavy minded people began to design ways to survive and thrive. They dared to build another Tower of Babel and reach heaven's gate for that is the way they felt the land of God whenever they desired. That's all they desired in their minds and thoughts. God never took the thought out of their minds and they transferred it to successive generations.

Man desired to know the Lord all the time because that is how he is created. If you don't believe in God then you have to believe in something. Man can not exist in isolation of faith because it is an integral component of his being. No man can live in absolute faithlessness before he dies and that is the law of the Lord. Anyone who believes in idols is only fulfilling that craving that is part and parcel of his mind. The mind searches and searches until it finds something to revere and do the dictates of the spiritual power. The mind that does not find a spiritual power simply dies and resents its existence. Anything that is believable to the mind becomes the man's god.

Thirty thousand of those who left Babel settled in Mesopotamia. Mesopotamia became a great nation inhabited by a very delightful people. Messages of the Lord went into the people's hearts and guided them in their daily lives. Knowing that many of them desired to build great monuments to honor the Lord. Statues of angels were erected everywhere in places of high concentration of people. And the Lord didn't mind knowing of their good intentions. Hatred never prevailed among the people in Mesopotamia. Man became enamored with the Lord and delighted in his ultimate powers. Every man knew the Lord and the Lord knew the feelings of their hearts.

After the flood that exterminated the areas where humans inhabited the Lord vowed never to abandon man and his spirit dwelt among mankind. Never will the Lord reject mankind in its totality and he saw the damage he did when he flooded the areas inhabited by man. However, the Lord was not in full control of every mind. Satan, the devil and demons consistently desired to sway the minds away from the ways of the Lord. The Lord wouldn't destroy them because he gave his word to give Satan dominion over hell and liberty to harvest followers from the earth. Knowing that he desired to send mankind a savior that will deliver them out of the clutches of the wicked spirits. However, no one was holy enough to deliver the people out of the wickedness of Satan and his associates.

Knowing that, the Lord was totally helpless and felt a total failure. Dirty minded people will feel such a notion to be an abomination. However, God feels very distressed sometimes. He has no desires to violate his words that were said earlier and has to live by them. No one should doubt the infinite powers of God and his ability to create and wipe out as he desires and wishes. That is never a question that should be entertained by the mind of any man.

Great things began to take place among the Mesopotamians and never forgot the Tower of Babel. Heaven was the desire and the thought of every Mesopotamian and knew they could find it by building a tower. Many found messages in their minds urging them to build a tower and that was the desire of Satan. Knowing that the Lord decided to disperse the Mesopotamians. There was a great mudslide that destroyed a large segment of the main city and wiped out the power of the Mesopotamian empire. The people desired no more to rebuild the city but instead chose to migrate to new lands and establish settlements. As a result, heavy migration took place and Mesopotamia was nearly empty of its inhabitants.

Every one chose different directions and travelled in small and large convoys to search for new settlements. Having said that, not many of them travelled great distances and found settlements not too distant from Mesopotamia. Many waited for a long time before they made their decisions and it wasn't a sudden evacuation of Mesopotamia. The thinning of the population took more than two hundred years. During that period of time Mesopotamia never recovered and was never restored to its original height. Those immigrants who desired to move in large convoys planned for many years before they departed and were well prepared for their journeys. Every migrant made enough preparation and acquired livestock, grain and methods of transportation. Never were there a people that starved to death throughout the entire migration.

Anything became possible for the Mesopotamians people. They settled in other areas and desired to develop images of man to represent God and knelt down and worshipped it. Even more, none of them ever forgot the Tower of Babel and their desire to go to heaven whenever they pleased. Wherever they went they attempted to establish a tower, no matter how small, as a replica to the Tower of Babel. They never died when they reached the summits of their little towers but they felt fulfilled every time they did so. That was the perception they got in their hearts. Every migrant was never satisfied until a tower was built in their vicinity. Everything they did was based on that. Heaven never left their minds.

Never were those who migrated desiring to deliver themselves to Satan and Satan was extremely frustrated. Demons constantly invaded and embedded themselves in the periphery of the hearts of the people but they were always desiring to get out because the souls of the migrants rejected them in totality. Only the devil was successful. The devil messaged the hearts of the migrants and began to sacrifice animals to appease the Lord. And the Lord saw that and never felt anything wrong with that. However, deadly acts began to take place among the migrants. Many began to sacrifice their children to appease the Lord and the God was disgusted.

God knew it was the handiwork of the devil. The people waited for a signal from their chief priests to delve into such practices. Usually a child with no blemish on his body was requested by the devil and searched among their children and delivered that child for sacrifice. Knowing that, many families began to give tribal marks on their faces to save them from being sacrificed. That is how defacing a child's face became the norm in many African nations. The face was chosen so that it would become very clear for the witchdoctor to look for other candidates. Eventually, various facial marks would be synonymous with the identities of different tribes. The devil requires the sacrifice of unblemished child because he feels children with scars on them are incomplete. He desires a complete child.

The Lord was mad and lashed the devil for his evil deeds but the devil told God that he was within his bounds and his authorized zone. God had restored a third of the devil's angelic powers and assured him protection to exercise his influence if he could succeed. The devil was banished from heaven because he thought of hurting other angels and that remained his ultimate desire even after he was thrown out of heaven and relegated to hell. The devil sought vengeance and found convenient victims in children because they have angels in their hearts when they are young.

Satan became sickly disgusted whenever he heard the traumatized angels in sacrificed children. And so were demons. God desired to delete the devil to eternity but was restrained because the devil worshipped God. The only reason why he was thrown out of heaven and condemned was because of his desire to keep tormenting angels and the devil never stopped worshipping God in heaven and thereafter. Demons worship God also and were in such a state because they became liars, excelled in deception, crooks etc. and were great worshippers of God in heaven and thereafter. Satan believes in God but never worships the Lord because he was thrown out of heaven because of that. The devil's ultimate joy is when he sees and causes the torment of an angel. The devil causes massive mayhem among men but that is not his priority and derived very little satisfaction from that.

Everyone that delighted in the devil and did his evil deeds never saw anything but hell. The devil didn't care if they went to hell or heaven. That is not his problem. Only Satan recruits souls to increase the size of his domain, hell. The devil never gave liberty to his follower on earth and dictated various laws to guide them. Satan has no laws formulated to guide those who reject God and join his close-knit membership on earth. On the contrary, those who reject God become absolutely free and liberated minds and live their lives independent of God's bondage and never feel any burden upon them. Once Satan sways one's mind to reject God he never ever interferes in his/her life and makes the infidel believe that he lives a perfect life for he knows his/her final destination when they die. Hence, those who follow the temptations of demons and Satan and reject God sacrifice their existence for a few years of ultimate liberty on earth and eternal torment, torture and bondage in hell when they die.

Still the ultimate decision to make a choice is entirely in the hands of man and God and the angels never interfere in such an important decision in a person's life. Even infidels have faith in something other than God. Having faith in God is an obsession that validates and keeps the soul alive. Infidels must also have an obsession in something. No Human can exist ever in a vacuum and without a single obsession. If man is void of a single obsession the soul fades and dies eventually forever. Obsession is what keeps the soul alive and worthy. If no single obsession is present in a man he is as good as a stone and ceases to exist. There are millions of obsession an infidel could choose from but he has to believe in at least one of them for his soul to remain alive. Unlike the faithful an infidel could have an obsession in a hundred things simultaneously that he holds dear in his/her heart. For example, those who believe in the theory of the evolution of man become obsessed with Darwinism and Darwin is their god. They don't know it.

An embryo also develops an obsession in his mother's womb. At the age of three months an embryo begins to receive messages from his mother's mind. The embryo and the mother's mind are interlinked with each other. The mother could perceive silent messages from the child inside her. At three months the embryo only perceive only important milestones in the mother's mind like extreme joy, extreme sadness. At the age of five months the embryo develops direct link with the mother's mind and perceives every notion in her mind. The embryo develops obsession at an early age and has love for the mother and she becomes his god. An embryo absorbs every notion in the mother's mind and stores it in his/her memory and never disappear for life and ultimately become the guiding principles of the individual. A baby never remembers what he/she accumulated in their memory during the incubation period but such notions become manifest at a much older age, some at twenty, some at thirty and some even at fifty years old. That is why a person has some characteristics of the parent at a later age.

A mother that loves her husband dearly gives a particular message to the embryo. The unborn baby knows before birth that the mother loves the father and has a favorable perception about him. When the child is born he develops extreme love for the father and tries to emulate him even if he still loves his mother and ultimately develops the characteristics of the father in an older age. A mother that resents her husband also sends a negative message to the mind of the embryo and develops hatred for the father in the womb and after birth. Such fathers can never have the love of the child born under that circumstance. Ultimately, the child develops the characteristics of the mother only at a later age. Those parents with normal family relationship with no overt love or hatred have a child that has loves for both parents. Such a child will develop the combination characteristics of both parents at a later age and will have equal love for both parents throughout their lives.

Miscarriages in women are nothing but as a result of thin placentas. As women grow older the placentas that regenerate during pregnancy are extremely thin and disintegrate easily. Some last a few months and others last longer but can not have the thickness to carry the embryo through nine months of pregnancy. Stillborn babies are not cursed children and have no spiritual associations. The mother as usual transmits the thoughts of her mind to the mind of the baby inside her. The baby begins to love what he hears from the mother and will want to stay inside her indefinitely.

At near-birth the baby becomes frantic and would not want to come out. He becomes helpless because he could not prevent his birth. As he gets closer to come out his heart begins to beat faster and faster. Eventually his delicate heart gives in and dies of a heart attack just before birth. Usually loving mothers experience such a tragedy. The child perceives extreme love and does not want to leave the mother's womb. Mothers should always perceive in their minds and tell the unborn baby that it is time to get out and that his time in the womb is over.

Before the age of a maximum of thirteen years the angel in a child's heart determines the time of death for the adult that the child will become. However, that is not the final date. The angel monitors the child until the age of thirty, which is the cut-off date. The angel then determines millions of factors including the choice the person makes in life and comes out with a specific date time and second in which he will expire. By the age of thirty the body develops into its maximum height and no change occurs in the composition of the body after that age. The person will live the rest of his life with the organs and tissues developed up till the age of thirty. After that age only deterioration takes place. The angel knows precisely how long every tissue and organ would last and which organ will ultimately becomes the cause of death of the individual. Hence, the date of death is revised at the age of thirty and it is the final one. For example if an African lives in the continent for thirty years and migrates to a country with the best medical facility and better diet his life span wouldn't change even by a second. What he acquired at the age of thirty is what will remain in him for the rest of his life. Even experienced doctors could sometimes determine a person's life span after considering all his health conditions even if not to the last second. In the near future doctors will be able to forecast a person's life span to the month and then even get better. Every human being has his own unique development of tissues and organs and that is why people die at varied ages. No human being had or will have the same composition to another, ever.

Humans are made to look like God because he wanted the same image and mind of his replica. Nothing is done without God's knowledge mainly because the human mind imagines in the same manner as the Lord. No animal interacts with human socially or otherwise because they don't share the same mind. God doesn't know what animals perceive even if he has the capacity to do so. Humans relate to each other mainly because they have the same storage capacity in their memory as their fellow man. Everything stored in the memory of the brain is stored in the mind simultaneously. The mind is the engine of the body. Without the mind the brain is just a worthless mass of tissues.

The mind messages the brain memory at the rate of ten times the speed of light and gets a response from the brain at the rate of a hundred times of the speed of light. When an idea emanates from the mind it is transferred to the relevant part of the brain instantly. The relevant brain part dissects and immediately analyses the idea from the mind and its search engine scours the entire memory and finds the stored message and decides the ultimate solution and immediately sends it back to the mind. That becomes a piece of knowledge stored permanently in the mind. That is how the mind becomes knowledgeable. The brain has no knowledge whatsoever. The brain is a storage of tidbits and tens of millions of varied information.

The brain doesn't have a complete knowledge of anything. For example if the mind searches for an image of a bird and seeks the help of the brain, the brain sees ten thousand messages in the brain that make up the composition of the particular bird. The brain searches the entire memory in less that a thousand times the speed of light. When the image of the bird is complete the brain sees a complete image of the bird and instantly transmits it to the mind. The mind sees a complete composition of the bird and stores it in its storage bank. From then onward when the mind desires the information of that same bird again it sends it to the brain and brain responds instantaneously and sends back to the mind because it has a memory stored of the complete image of the bird already. It doesn't restart its search again to form the image of the bird again as it did in the beginning when the mind and brain did not have a complete image of the bird. The complete image of the bird, in its entirety, in stored in the both the memory of the brain and the storage bank of the mind.

The mind doesn't know that the complete image of the bird is stored in its bank and the brain doesn't know that the complete image of the bird is stored in its memory. That is why men don't have the image of the bird all the time in their minds and the brain doesn't send the image of the bird to the mind all the tile. Hence, the desire to see the bird by the mind becomes an entirely new message perceived by the mind as if it has never ever imagined it. The message from the mind goes to the brain and its memory searches and finds the image of the complete of the bird very, very fast and transmits it back to the mind. The mind feels stupid when it discovers that it had the image of the complete bird in its storage bank already. That is how people recollect past events. Once the mind finds the image of the complete bird in its storage bank a ripple effect takes place. The mind accesses every aspect of the bird in its storage bank without consulting the brain. That would be the total knowledge the person has about that particular bird.

The mind doesn't care and constantly accesses the brain all the time. Anytime that happens the same result occurs and the mind increases in knowledge all the time. Many minds that desire to access the brain more frequently end up as geniuses. A genius doesn't have a larger brain than other members of humanity nor the color of its brain matters. It is purely determined by the number of times a genius's mind accesses the brain. A genius finds it delighting to search his mind all the time, more than other human beings, and the brain is bombarded with deadly numbers of information requests from the mind.

Many geniuses develop paranoia and become seclusionists because their brain is constantly processing data continuously and without stop even while they sleep. The mind stores so much information, many of which is not relevant to the existence of the genius, and the mind is clogged with so much storage and that weighs so much in the life of the genius. That is why geniuses stay in seclusion because they realize that they have no knowledge to gain by interacting with people. They look down upon normal people. Their separation from their fellow man develops into paranoia because man is created to associate with his fellow man for his mind to remain intact. That is even relevant to animals.

The genius delights in his imagination and would not exchange it for anything in the world. The mind of a genius is mainly very different from a normal human being because it can transmit and seek help from the brain ten times more ideas than a normal human being. That is why a genius could answer a question much faster and elaborately far more than their normal counterparts. That same response would take far more time to figure out by the mind of ordinary people. Every man has the same power for the number of ideas the mind could send for analysis and results to the brain.

Many geniuses exceed that limit and the mind is stretched beyond its maximum capacity to send ideas to the brain to retrieve all the relevant information from its memory. That state of mind is a grave danger to the genius. The genius becomes fired by sickly decisions and becomes delusional. His delusion has nothing to do with good or evil spirits. It is purely as a result of a clogged mind. The genius's mind doesn't disintegrate or loose any of its capability even if it is highly clogged with constant ideas. The only thing is that the processed responses from the brain become fragmented and the mind knows distorted messages. Mad scientists are as a result of a very fine mind that has lost the ability to receive more completed messages from the brain after processing the feeble ideas emanating from a gravely and sickly mind of a genius.

Delusional disorder is the disease of the mind and is not as a result of a defect in the brain. The brain of a delusional person is in perfect condition and desires to process ideas transmitted to the brain from the person's mind. No damage is done to the mind and brain relationships. People with delusional disorder are victims of demons. Demons whisper really evil messages from the periphery of the person's mind. The person begins to believe the demonic whispers as if emanating from his mind. Gradually, he begins to see images of the ideas imparted in him by the silly demons. The mind sends the name of an angel to the brain and the brain process it and sends back a complete image of an angel that was already stored in its memory for the person had heard and figured out the shape of an angel before. The mind finds it in its storage bank and the person realizes the exact image of an angel.

However, when the person sees the image of an angel in his mind the demon inside him fights him desperately. The demon doesn't want him to imagine anything about angels. However, the person persists in his mind. The demon doesn't just agree with him but also whispers and convinces the person that he is indeed the angel he sees. That is how delusional disorder afflicts a person. The demon imparts all kinds of spirits in the mind of a person that falls victim to the deadly messages of demons. Some think they are messengers of God, some think they are devil's messengers, some see themselves as super humans with supernatural powers. Even some believe they are God with supreme powers to change the universe.

Healing a person with delusional disorder with medication works marvelously because it deadens the state of the mind while on regular medication. Persons with such a disorder make rapid progress when treated by a psychiatrist and delight in their skewed messages no more. Barely ten percent relapse and become victims to demons again. When the demon realizes that the mind of the inflicted person is not responding to its whispers it is highly disgusted. It tries aggressively to maintain its influence in the mind of the disturbed person.

However, the medication reduces drastically the sick person's ability to seek information from the brain for things that are not relevant to his daily livelihood. The demon becomes frustrated and finally disappears. All those suffering from delusional disorder recover fully and there remains no residue of their illness ever again unless they are inflicted by demons again which is as likely as being struck by lightning. Those who relapse are only those who still have demon in them and only remained dormant for sometime. When fully recovered they are mostly appalled by what they thought of themselves when they were controlled by demons.

Schizophrenia is a different story. Schizophrenic people are not devil possessed or influenced by any external spirits of any kind. Schizophrenia is just a simple and tiny rewiring of the brain. Nothing more. People with schizophrenia don't have serious brain damage and the single damage of the brain doesn't not deteriorate and expand. It is a one-time change and does not repeat itself.

People with schizophrenia are usually very good listeners and is their primary characteristic. This doesn't imply that all good listeners would develop schizophrenia. Far from that. All good listeners store relevant information better than casual people in their brains. The has no different functions in a schizophrenic person as safe people. The mind of a person perceives and reacts to different words differently. When a person hears what he perceives to be good words he reacts casually. But when he hears harsh words the mind and the memory of the brain give prominence to such words. The brain is highly compartmentalized. It has nearly two million compartments in the memory of the brain and there is no such compartmentalization in the mind. All ideas are given equal meanings and equal spaces in the storage bank. Hence, the brain has special compartment for good words and another for bad words. The bad words are by far more prominent in the memory because of the person's reactions when they were heard and the impact they had on the person.

Schizophrenic persons are people who suffered a deadly and terrible headache in their lifetime. It is a one-time experience. It is usually a fever with about two hundred degrees and nearly dilapidated the person when it occurred. The fever lasted probably for two to five days but the high temperature probably for two or three hours before it was controlled and reduced to a tolerable level. At such a critical period the brain desires to devise a way to fight the deadly temperature and attempts to restructure itself. Noting that the brain finds a simple way. It diverts some of its memory messages to the less condensed part of the memory. It assumes that the large storage of information concentrated in very small part of the memory is responsible for the high temperature. It is a fact. Memory concentration is always the cause for headaches. Headaches disappear when concentrations of memory disperse throughout the memory parts of the brain. Medications deaden the nerves that transmit the pains but do not heal the root cause of the headache.

When the high concentration of memory stored in a small section of the memory part of the brain desire to disperse they are hampered by the high temperature. Normal headaches are resolved easily and the dispersion of the concentrated memory information take from a few minutes to two or three hours. Those who suffer persistent headaches are just people who disperse their concentrated memory information to just alleviate the pain. They don't totally disperse and hence, some sections of the memory are quickly filled up and concentrated and the person develops headache intermittently and within a short span of time. Those people with a high concentration of memory and achieve total dispersion into various parts of the memory part of the brain never experience headaches for a very long time and probably never. The dispersed memory information is spread over a large parts of the memory brain and it will take a long time before any part of the memory achieves the high concentration that results in headaches.

A person with schizophrenia does not achieve the dispersion of the highly concentrated part of the memory. The memory becomes desperate and wants to disperse the concentrated information to other less concentrated or vacant parts of the memory brain. It doesn't succeed. Hence, it devises other outlets to circumvent the normal routes of dispersion and attains access to the less concentrated part of the memory brain through these passages. The person successfully recovers from his headache and is excited. But shortly he begins to feel funny things in his memory. The abnormal routes created to disperse the high concentration of memory information never relax. They continue to transfer memory information to other parts of the memory part of the brain. Eventually, it safely transfers all the memory information from one part of the memory brain to another and entirely new location. This does not produce the slightest change in the normal functions of the brain and its processing ability or its link with the mind. It is just a change of location. Period.

When the change of location is finally accomplished the abnormal routes desire to remain active but will have no information to transfer. Suddenly a decision is arrived and the abnormal routes will desire to make more information so that they will transfer it to the new location of the memory part of the brain. But there will be no information because new information perceived by the person go directly to the new location and are no more stored in the old locations of the memory. Hence, the abnormal routes have no functions and values. But they don't see themselves as such. They believe that they are an integral part of the memory part of the brain and devise functions for themselves to ascertain their relevance.

The abnormal routes will desire the memories stored in the new locations to return to their old places through them again. But the part of the brain that controls the functions of the memory part of the brain does not see the relevance because the person does not experience a terrible headache. Moreover, the memory information in the new locations are made to ensure they are dispersed over a large segment of the memory brain to avoid the reoccurrence of the tragic circumstance that was responsible for the creation of the abnormal routes. But the abnormal routes persist and the message gets to the part of the brain that controls the functions of the memory part of the brain. The main brain permits mainly the prominent stored memory information to go back to the old location of the memory. The main brain that controls the functions of the memory parts of the brain searches for such information in the stored memories and locates the bad words compartment to be the most prominent part of the stored memory. Hence, the abnormal routes are excited because they are now relevant as integral components of the memory part of the brain.

The bad words compartment migrates back to the old location of the memory until it is finally relocated and stored in the old location from where they initially moved. Still no visible change is felt by the patient only slight irritations that are irrelevant. Nothing in his head changes and nothing obvious signs. The deadly routes again desire to send more memories from the new locations to the old location as they did with the bad words compartment. But the main brain that controls the functions of the memory part of the brain would not find prominent information in the new locations that desired to go back to their old locations. They become comfortable with their new location and the search engine accesses them just as when they were in their old location. The bad word compartment memory were glad to go back to their old location because they were disgruntled knowing that the patient never accessed them that much. People access the good words memory compartment often and rarely delve in the negative aspects of their lives. As a result, the bad words compartment knew that it would release its information easily if it was located separately.

So the abnormal routes become redundant again but now have no control over the bad words compartment. The bad words compartment memory are just happy in their old location and have no desire to go back with the rest of the memory compartments. The abnormal routes desire to become more valuable and devise another suggestion to the main brain that controls the functions of the memory part of the brain. They will suggest and say, let the information stored in the bad words compartment be transmitted to other parts of the main brain that controls the entire functions of the brain in its totality. The main brain resists vehemently and rejects the suggestion by the abnormal routes. The abnormal routes persistently send vibrations to the bad words compartment of the memory brain and desire them to send similar suggestions to the main brain. The Bad words compartment deletes the sickly suggestions of the abnormal routes and discards them. However, the abnormal routes permit them to suggest a way out for the redundant and less accessibility of the bad words compartment by the search engine of the memory brain for the patient hardly remembered any of them entirely.

The bad word compartment of the memory suggest that the best way for them is to transmit their stored information to the center of the body. The bad routes knew the impossibility of such a desire and told the bad words compartment that their suggestion was not attainable. But the bad words compartment never, never ever gave up. Eventually, the main brain made an alternative. Instead of the bad words being accessed by the search engine of the memory part of the brain a different approach was made. The part of the memory part of the brain was deleted and newly created memory was established in the audio part of the main brain. The new bad words memory was now constantly accessed by the audio scanner and released to the membranes responsible for hearing. The person suddenly heard very bad words as if they were spoken by a person speaking to him through his ears.

The person inflicted with such dirty voices can never know where he heard the filthy words even if he attempts. If he tries to imagine why he is constantly called an idiot in the voice he hears and searches his mind he will never ever find out. His mind will want to know who and where he heard such a word in his life time. His mind sends the idea to the memory part of the brain to be processed and find the solution. The memory part of the brain scans and searches its entire compartments and will find no such record in its storage because the bad words compartment was deleted from the memory part of the brain when new memory was created in the audio part of the main brain. The mind has no link to the newly created audio memory neither does the main memory part of the brain. The person has no control of the audio memory because he can only access it through his mind. As a result the audio memory transmits its information independent of any control and does so repeatedly and relentlessly. That is why schizophrenias hear voices constantly, day and night.

Schizophrenias should know that the vile words they constantly hear are nothing but words they heard before. They hear no new word they never heard before. They are just prominent words that they stored in their memories from early childhood. Many of the words could be from movies, parents, friends or just curses they heard on the streets. All the words they hear repeatedly were heard by them at one time of their lives or another. Schizophrenias should delight in the words instead of being despondent. Some of the words they hear could be traced to a particular person and place. If they had a cursing father, mother or friend. Sadly, the voices they hear are brain voices unique to schizophrenic patients. They can not identify the voice they hear as that of the mother, father or friend because they hear the voices in a single frequency and tone. But they can easily figure out when they precisely heard the bad words because they had an impact on them when they initially heard them and that is why they were stored as prominent words in their memory.

The audio memory remains forever and is never deleted. The audio memory has no impact in the functions of the audio system of the body whatsoever. All schizophrenic have no problems with their hearings and only listen to daily bombardment of sickly and despicable words throughout their lives. No schizophrenic should be disturbed by the voices they hear. They should learn to enjoy them and see themselves as unique and special people. Hearing voices should not impact their daily lives and should accept them as they see the rays of the sun. They should laugh when they remember the circumstance under which they heard the bad words. Schizophrenias could never ever be healed permanently. Medications could only reduce the tones slightly. Their brain is slightly rewired and there is no way to restore the original form of the brain. As for the abnormal routes they will eventually die down when their functions eventually become absolutely irrelevant.

A Bipolar mainly remains absorbed in facts of his head. A bipolar has no changes in the compositions of his brain. He never loses any of the powers of his memory and neither does he the discerning powers of his brain. A bipolar remains active throughout his disease manifestation and doesn't rely on the devil, Satan or demons. It has nothing to do with them. Many bipolars live a perfectly normal life and don't exhibit any abnormalcy in behavior. Knowing that a bipolar has a sickness that is not treatable and lingers for life. Normal change of behavior is required to cope with the severity of the disease. Knowing that many bipolar could achieve success and live with their abnormalcy contained to a minimal level of tolerance. Heavy dosage of medication can alleviate the disease but a bipolar could live a perfectly normal life with feeding him the lethal and poisonous medication that is prescribed for him by trained psychiatrists. Most medications consumed by bipolars contain deadly combinations and the side effects are dangerous to the diseased person's existence. Some develop diabetics, some heart murmuring and others kidney and liver failures.

A bipolar develops his sickness when silly things happened in the memory parts of the brain. Some members of the memory cells become severely unused and begin to deceive themselves. The members of this disgruntled cells begin to hate one another and delete one another. They eventually end up dead. The reason why they die is because their abnormal behaviors are transmitted to the main brain that controls the functions of all the other segments of the brain. The main brain discerns their messages and discovers that they are a grave threat to the very existence of the entire memory if they were allowed to multiply. Hence, the main brain sends daily warnings to tell to change their abnormal demand. Their abnormal demands are, among others, to discern messages that are outside the memory of the person to ascertain their values and release messages. That become an absolute no no. The main brain realizes that the abnormal cells are a threat to the existence of the entire brain system and deletes them before they multiply

The dead cells remain in the memory and become a different category and with their own separate compartment. The brain scans the memory when an idea is received from the mind of the person with the dead memory cells. The search engine goes through the entire memory to retrieve all the required relevant information to respond to the idea transmitted by the mind. Suddenly the search engine scans the dead cells compartment also. The search engine goes through the entire dead cells and retrieves no information. However, the scanner does not proceed to the next memory compartment and continues to search unsuccessfully to retrieve information from the dead cells. The scanner doesn't know that the cells are dead and the main brain did not see the relevance to inform the search engine of the memory. Hence, one memory strand become stuck for a period of time and returns over and over and over to go through the dead memory cells.

At the end the memory compiles its search and sends a response to the mind but without the information from the dead cells. Even though the dead cells contained no information before they were killed by the main brain they remain relevant because the final conclusion of the memory must be reached after going through the entire compartments of the memory. Hence, the transmitted message to the mind from the memory would have a flaw and the mind rejects it. The reason why the mind rejects it is because when it receives the completed message from the brain it tries to access its vast storage bank it does not locate the image sent from the memory and would not know the memory sent message. For example a bipolar sees a bird and the eye memory sends the message to the mind.

The mind gets the message from the eye and sends it instantly to the memory part of the brain and the memory scans the brain and compiles the image of the bird and sends back to the mind for recognition and ultimate storage in its storage bank. And the mind receives the compiles of the image of the bird and searches its storage and discovers the image of the bird previously stored in its storage bank. Hence, it realizes that it is a robin. Automatically when the mind identifies the object as a robin all the stored messages related to robins are revealed to the mind. That is why you know all the details you had previously known about robins immediately you see one.

It has to be information known or stored in the storage bank of the mind. One person may know that robins sing while would have no clue of such a fact because that information is not stored in the storage bank of his mind.

A bipolar sees a bird and the process goes as explained above until the final image of the compiled image of the bird returns to the mind for final identification of the bird. The mind scans its storage bank but surprisingly doesn't find any storage of such image in its storage bank. The scanner of the mind sees a very, very close replica it has of a robin but knows it is not the same bird. Hence, the mind stores the new image of the robin and categorizes it as an unknown category because the mind doesn't know anything of the new image of the robin. Ultimately the only record it would have about the new image of the robin is exclusively its image and absolutely nothing else.

Many scanners are available in different parts of the entire brain unique to the various sections of the brain. The memory has more than two hundred thousand scanners and every one of the scanners is relevant in determining the final conclusions of the memory that is sent back to the mind. The image of the robin is the information gathered from the scanning of two hundred of them. When dead cells occur the single scanner that searches information from them becomes redundant for sometime. The single scanner just goes on and on scanning the dead cells thinking that they have some relevant information for the formation of the image of the robin. Helplessly the single strand of memory deletes some of the information it had gathered before it began to scan the dead cells.

Eventually the single stand of scanner that was stuck on the dead cells abandons its relentless effort and contributes its gathered information to the compilation of the image of the robin. But the final information gathered by the single strand that was stuck in scanning the dead cells is not complete and defective because it deleted some relevant information it had gathered before it began to scan the dead cells. As a result, the final image sent to the mind by the memory is defective and is not an absolutely a replica of the image of a real image of a robin stored in the storage bank of the mind previously. Hence it becomes a new image in the storage of the storage bank of the mind and categorized as unknown image. Only bipolars have an unknown image compartment in the storage bank of the mind. No other human beings have such a category.

The defective part of the image of the robin seen by a bipolar is one trillionth part of the entire image of the robin. The mind must receive an absolutely perfect image of the robin to identify the real robin in its bank and know all the relevant information about the bird instantaneously. So the bipolar sees a robin but has no clue what type of bird it is for sometime. That duration could vary from a few seconds to a maximum of five minutes in severe bipolars. But there is a memory lapse in bipolars regardless for how long it lasts. And the reason for the varied duration is because the mind sends many, many messages to the memory attempting to know more about the new robin and store it in its storage bank. However, the memory finds blank information about the new robin and only sends back the same image of the robin with no additional details because the patient never messages additional information about the new robin image.

The mind never stops to know the details about the bird because the patient's mind demands it to complete its final complete relevant information about the robin. And this back and forth goes on and on and on for millions of times. At the end, the single scanner learns to skip the dead cells and scans the rest of the compartments it is assigned with. This is the period of paranoia in a bipolar.

Suddenly the mind gets the complete image of the robin that is a replica of the original robin in its bank and instantly knows all the details that were stored in its bank and the bipolar recognizes the bird as a robin and all the relevant information about it. From then onward every time the bipolar sees a robin the process to identify it takes place instantaneously but this time the single scanner does a normal scanning activities and no more gets stuck when it scans the dead cells. But this fact is only true with the image of robins alone. What ever image the bipolar sees it has to go through the same process the mind and memory went through for him to identify the robin. It has to be dome mainly one-by-one. Hence, the memory of a bipolar discerns things very slowly and thinks the devil is doing to him/her. Only images and what the bipolar sees by his eyes are affected in the defect of a bipolar and all the other senses are unaffected.

The bipolar becomes despondent when his mind doesn't discern in the gathering of details at the normal rate of a person. He tells when the scanner is stuck on the dead cells. His head pains because the information lost by the single scanner strand heavily tamper with the true functions of the memory. And deadly messages could be imparted and the sad, paranoid and defective mind of the bipolar could reduce him to a helpless wreck during an episode when the deleted messages hamper the safe functioning of the brain.

Information in the memory are never deleted and last until the death of the person. However, in a bipolar, the single scanner searching the dead cells deletes some of the information it had gathered earlier before it began to scan the dead cells thinking that they are the real reason for its inability to scan and retrieve information from the dead cells. The information the single scanner deletes does not disappear from the memory and could be retrieved again by any scanner.

The only thing is that the deleted information remain within the memory outside their compartment from where they were retrieved and never go back to the same category to provide the same information they were delivering. The deleted information have no same functions as the other members of the memory. They become deadly disruptive in the normal functions of the memory. Bipolars know everything they want to know. None of them is overtly evil as some schizophrenic patients who act on the instructions of the voice they constantly hear. Bipolars know when they are normal and when they go to remission. No one could heal a bipolar. It is a permanent defect. Knowing that bipolars could wean away the time of the disruption of the deleted messages in their memory.

The deleted messages do not delete any memory information and hence, the memory of a bipolar is always intact. However, for sometime the disruption of deleted cells could be serious and last for sometime. No bipolar could determine how much disruption will take place in their memory and for how long. Some disruptions take place for a few minutes and others go even for more than two days. There is nothing they can do about it. Medication never ever works and no one could remove the disruptive information in a bipolar's memory. As the process of identifying an object goes through the process of identifying the robin more and more deleted information become available in the memory until the bipolar exhausts identifying all the relevant objects he sees in his daily life. That is why bipolar defects become stationary and do not deteriorate after their stable condition is reached.

A bipolar loses some sense of his reasoning during an attack. He goes on and off depending on the disruption of the memory by the deleted messages and the period of time it lasts. When the deleted messages become fairly dispersed the bipolar shows normal tendencies and amazing emotions. During an attack his mind feels like he has cloudy messages. Nothing detrimental in being a bipolar. Knowing that bipolars should learn to cope with their illnesses. They should know that the development in their development is a passing phase and would not last forever. Many bipolars become very despondent and really depressed to the extent of committing suicide. Now they know the real reasons for their defect.

The bipolar knows that he is demented and assumes he has brain damage. Nothing could be further than that. Nothing in the brain is damaged except the nearly two hundred cells that permanently die. Having said that the bipolar thinks his head is screwed up and has deleted memory. A man married to a loving wife believes differently when he is inflicted with the disease. Hatred results at the end. When a man's decision mechanism is altered the decision he makes after that attack is silly and does not reflect the actual scenery. The man who loves his wife dearly doesn't recognize her when the scanned of the search engine is stuck on the dead cells.

The duration it lasts may sound insignificant but has an adverse impact in the man's relationship with the woman he loved. During the entire period of not recognizing the image he sees as that of his wife his memory is imprinted with that image. When the signal of the wife's appearance is sent by the optic nerves to the mind, the minds sends it to the memory of the brain for analysis and final results. The memory of the brain scans the entire brain memory and a single scanner gets stuck on the dead cells. This single scanner then deletes some of the information it had gathered before being stuck on the dead cells.

The deleted messages become like a debris in space and end up as obstructions to the smooth functioning of the memory part of the brain. At the end the single scanner abandons its search of the dead cells and delivers its messages to contribute in determining the image and identity of the woman that the man was seeing. The completed final result returns to the mind with a message of a different image of a woman. The reason is because the messages of the single scanner was defective and incomplete because it deleted very relevant information it had gathered before it started scanning the dead cells of the memory.

Hence, the memory of the brain and the storage bank of the mind store the new image of the wife. When the mind searched its memory bank to find out all the details about the wife the man he was seeing comes out with absolute blank. The memory of the mind has detailed memory of the wife's image but not even a single message of the newly stored incomplete image of the wife. As a result, the husband sees a silly message. Even though the man sees a complete image of his wife his mind memory has no record of her in his mind. There is no message of the supposedly distorted image of the wife in the memory bank of the mind because of the deletion of some relevant information about her during the scanning of the single strand's search of the dead cells. As a result, the husband sees the newly formed woman that was compiled by assembling the various search engines results of the memory including that gathered by the seriously deficient single strand that got stuck on the dead cells and had to delete some of the information it had gathered before it reached a dead end when it tried to retrieve information from the dead cells.

The image the husband sees is alien to the man and never had seen her before because the mind has no record about her whatsoever and hence, the husband knows nothing about her. This may last for two seconds but is sufficient to place a doubt in the husband's belief. From then onward the husband knows that she could be someone else. Having said that the relationship deteriorates between the man and his wife. The wife would desire to know why he was rejecting her but he has no clue why he was doing so. He becomes delirious and this state mind is prominent only when he sees his wife. He still believes who his children are, his family members etc. Nothing else changes.

But the bipolar doesn't stop with the distortion of his wife's image which resulted in the condition above. Gradually the process continues and his children and family members become victims as well and would have doubt about them and desires to discard them. If this process continues indefinitely then the bipolar would loose touch with reality. Luckily it has its end. When the main brain discerns the messages of the brain memory it is alarmed by the near replicas of millions of images in its storage. It immediately wants to know the reason. Hence, a message is sent to the mind from the main brain and an idea is created. The new idea wants to know why there are numerous near replicas of numerous objects in the memory of the brain. The brain discerns it and sends its final findings and acknowledges the mind that all the near replicas have different messages gathered by that single search engine scanner.

The mind going through the process of finding an ultimate solution sends its directives to rectify the abnormalcy introduced by the single strand of the search engine. The mind sends a message prohibiting any search engine strand from accessing dead cells and limits the duration of time a scanning strand could spend on a particular section of the memory. As a result no distorted images are ever created even if more cells die in the husband's memory again.

Alzheimer patients are the most pitiful people suffering from any disease. They loose their minds and are made to die shortly thereafter. Man can easily solve the infliction of Alzheimer disease with a blink of an eye and stamp it out from humanity once and for all. Nothing is easier from the permanent solution to resolve the Alzheimer plague from infesting humanity.

A child with a perfectly normal body functions wakes up in the night and is alarmed by what he dreamt. The mother comes to the aid of her child and soothes him and he goes to sleep again. Knowing that the child never suffers and delights in subsequent night when he did not experience a similar nightmare. Over his childhood, the child wakes up five to six times in the night because of some form of nightmares. When the child grows older he totally forgets his childhood experience and his soul, mind, heart and body are never affected in any way until he dies. In his later days, the same child that grew into a man experiences some hallucinations a few times also. That also disappears and has no impact whatsoever in the functions of his entire body.

The symptoms of Alzheimer disease could begin as early as the age of thirty. Some start at forty, fifty, sixty and seventy and never begins above that age. Lets assume the adult man's Alzheimer symptoms manifested themselves at the age of fifty. At that age or any age that the disease manifests its symptoms first for any Alzheimer victim, the man is less stressed, mild mannered, and hence, less desirable to feed his mind with new ideas. These and other characteristics the man grows to develop are no symptoms of any disease at all. It is perfectly normal to develop such a state of mind as a person grows older, begins to be comfortable with himself and sees a brighter future and so on and so forth. And the mind responds to his gradually adopted condition and nothing happens at all. Perfectly healthy and perfect mind. Nothing to worry about.

The mind produces less ideas because the fifty-years old has no desire to deliver new messages. His curiosity and his surroundings become less important to him. He just takes everything for granted because he lived most of it and has adequate knowledge stored in his mind bank and brain memory. Again, nothing abnormal about that and perfectly normal. The mind become less and less busy as less and less ideas it creates. Remember that mind ideas are just response to stimuli generated by the different senses of the body, environment surrounding them and nothing else. For example, when a part of your body itches an idea is created the sensory nerves transmit the signal to the mind to determine what manner of action should be taken to solve the itching problem. The mind doesn't know what the itching signals are, just an idea, and hence, it sends it to the memory of the brain for assessment, analysis and final identification of the itching. The memory of the brain goes through all its storage of information and assembles are the findings gathered by the numerous strands of its search engines and its final conclusion is returned back to the mind.

The mind now realizes it is nothing serious and is just itching from the final findings of the brain memory. Hence, the mind scours its memory bank storage and gets all the relevant information connected with itching. What it is and how to resolve it. Hence, the mind sends signals to all the parts of the body required to scratch the spot where the idea of itching emanated. The part of the body is scratched by the hand and nails and the itching disappears. Problem solved. The reason why the mind recommended scratching the itching part of the body is because it had experienced such a problem before. As a result, it had the name of the idea and all the various methods a person could use to alleviate the itching problem. If the person had other alternatives of relieving itching, such as using cream, applying water or oil etc. all of them are stored in the memory of the mind storage bank. When the mind discovers it is just itching from the information of the memory of the brain, it automatically makes the various methods of resolving itching available and chooses anyone that is easier and handy to apply.

When something new becomes the stimuli the mind goes through the same process. The stimuli, for example flying saucers, are fed to the mind from the optical nerves it becomes an idea instantly. To discern the idea the mind sends it to the brain memory where it is dissected, analyzed and eventually built into an image of a flying saucer. The eye doesn't see a flying saucer. It just sees millions of trillions of light reflections at varied intensity that are reflected from the surface of the flying saucer. The mind initially receives these reflections and would have no clue as to what it is. It is the brain that assembles these reflection of lights into a complete flying saucer image. Hence, when the mind receives the image of the flying saucer it searches its storage bank for all the details on flying saucers but finds nothing.

And hence, the man realizes he was seeing a flying saucer, however, he would have no idea what it is for there is no additional information in his mind memory storage. The mind stores the new information in its memory bank as would the memory of the brain. The person's auditory senses hear the name of the object and the brain associates the shape with flying saucers and knows it as flying saucer. As more information of flying saucers is gathered from all the sensory organs they become new ideas and the process is repeated over and over between the memory of the brain and the mind. Soon, the mind and memory would have a wealth of information on flying saucers stored in the memory bank and the storage bank of the mind.

When the person sees another flying saucer the eye sends what it perceived, which becomes an idea, and is transmitted to the memory of the brain by the mind and the brain memory interprets the optical signal, dissects, analyzes and finally assemble the various components of the signals into an object and realizes it is a flying saucer because it has its replica stored in its memory. The completed flying saucer is sent to the mind. After the mind identifies the object it goes through its memory bank and retrieves all the information stored in its and knows as much as it is stored in its memory bank.

All what you read, hear, see about flying saucers go through the same process and are finally stored in the brain memory and mind storage bank. That is why when the man sees a flying saucer again he would have a wealth of knowledge about the object depending on how many ideas were created and dissected and rebuilt by the memory of the brain and stored in the mind's storage bank. Knowledge is data stored in the mind storage bank and not in the memory of the brain.

The brain has absolutely nothing to do with a person's knowledge and wisdom. All the different sections of the brain transmit their signals to the mind and the person has no perception of them. The signals are just ideas that become knowledge after they undergo though the process described above and finally stored in the mind memory bank for future retrieval. The brain memory has the largest segment of the entire brain. It is seventy percent of the size of the whole brain. The main brain controls the functions of all the different segments of the brain, including the memory brain. The main brain cells have simple structures compared to the complex structures of the memory brain.

To understand the sickness of Alzheimer one has to have a perfect knowledge and relationship between memory brain and the mind. If you don't understand fully that relationship you will never understand what goes on in the mind of an Alzheimer patient.

The man with a perfect mind and body lives his life to the fullest and in a comparably saner environment. Older people are prone to the disease because that is the age we attain peacefulness and our life is less rigorous and as a result create less ideas to be processed in the manner described above. That is perfectly normal and is not the cause of Alzheimer. And also not all, and only, people that portray such characteristics become victims of Alzheimer disease. The key indicator is that the mind receives less ideas and makes other decisions to occupy and keep itself busy. That is the main cause of Alzheimer.

When the mind is less busy and has less ideas to process it delights in going through its memory bank. The Alzheimer prone person remembers and imagines all the incidents in his life whenever the mind accesses that stored message in his memory. There is nothing wrong with that and is no indication of any illness. We all experience such things throughout our lives. As it scours its storage bank the mind discovers five or six abnormalities in its storage bank. It sees, for example, a cat, into different locations of its memory bank each with varied details of information. One cat has the details of a ferocious animal the size of an elephant while the other stored memory has the regular size and functions of a regular cat. The mind memory ponders and wants to find out why it so. The mind memory discovers that the giant and ferocious cat was the processing an idea that was triggered when the man was a baby and was traumatized by it once when he had the nightmare. And the regular cat was the idea started at a much older age. When this process is going on the man perceives how he used to be scared of a cat when he was young and now how he loves his pet cat.

The mind memory is disgusted because the giant and ferocious cat details in its storage bank is absolutely false and realizes how much damaging it was, and could be, to the wellbeing of the person it inhabits. As a result, the mind makes a decision and vows to reject bogus and damaging information that it would receive from the memory brain after assessing an idea and sends it back to the mind. Know the mechanism of converting an idea into useable information by the mind from the explanations above.

The mind begins to sift silly ideas processed by the brain and sent back to the mind for action. Whenever it receives such information it is rejected instantly and never stored in the mind memory bank. For example if the mind receives final information from the memory brain saying that "water is God" it rejects it instantly as a silly thought. Normally, when an idea is processed by the memory brain and the finalized information is returned to the mind it instantly scours its memory bank to know all the detail of the single information. But with the finalized information that it designates as silly it rejects it instantly and does not search its memory bank to know all the details about it. Such information fizzles and disappears and doesn't linger in the mind.

So the mind continues doing so for many years and no vital information is lost and the person remains in a perfect state of mind. Nothing lost because only irrelevant information are discarded. However, the mind begins to reject processed information from the brain memory that it considers less desirable in its memory bank. This is not a good idea because it begins to distort the man's perception and he becomes mad whenever he doesn't discern an issue to its fullest. For example, whenever the mind receives the processed image of a car from the brain memory it rejects it and the information fizzles and disappear, never to return ever. So the man would be very knowledgeable in every topic and issues in the world but gets lost when it comes to the issue of cars.

Knowing that no one should perceive that Alzheimer is as a result of frequent nightmares that manifest themselves at older age. Nothing could be further from the truth. The stored information in the mind memory that triggers the rejection of silly ideas doesn't have to be a nightmare. Any milestone or even fantasy entertained during childhood and before the age of thirteen cold be the cause. For example all of us imagined and believed many things such as horses could fly and many dangerous situations and avoided them. The brain of children continues to grow up till the age of thirteen when it finally stops and the person has to live with that size for the rest of his life. Before that children are fascinated, unknowingly, when their world begins to expand also and imagine a few fantastic and sometimes fearful things. Any of these bogus information could be the cause for the mind to begin to cleanup its storage selectivity.

So the permanent lose of the image of a car and all details of its information is the first symptom of Alzheimer. The person still sees a car and the optical nerves send the information to the mind. And the mind sends the optical information to the brain memory, because the mind doesn't have the capacity to decipher optical information. The brain memory dissects, analyzes and eventually forms the optical information into the image of a car and sends it back to the mind. The mind instantly knows that the optical idea was indeed that of a car. In normal instances the mind goes instantly to its memory bank and retrieves all the details pertaining to a car that is stored in its memory storage and the person would have full knowledge of a car. But since the mind has discarded car as an irrelevant information it ignores the information and does not go to its memory storage to find all the details about cars. As a result, the victim sees an image of the car but has no clue what it is. He knows details of every thing else but becomes blank with regard to cars.

So the mind begins to discard more and more information returned from the brain memory that were processed from an initial idea and soon the person loses the knowledge of many things. From the initial discarding of silly information to the first symptom of vital information could take up to ten years. It is not an instant deterioration. Eventually everything processed by the brain memory and sent to the mind is rejected and the man would have no knowledge in his mind and will be said that he lost his mind.

The Alzheimer victim could live like that for a long time if nothing happened. However, the main brain that controls the functions of all the different segments of the entire brain, including the brain memory, desires to intervene. Hence, it creates a new memory bank in it segment. From then onward ideas that trigger the process of the brain memory are sent from the mind and when the brain memory eventually comes with the ultimate result instead of going back to the mind it is diverted to the newly created memory bank in the main brain. For example when the man who has lost his mind sees a bird, the optical nerve still sends the information to the mind and the mind sends it to the brain memory. The brain memory does its functions and concludes that it is a bird. However, instead of sending the completed information back to the mind, which would have rejected it, sends it instead to the newly created memory in the main brain.

That same person sees a bird and the optical nerves send the idea to the mind and the mind sends the information to the brain memory and after the regular processing identifies it as a bird and instead of the final information going back to the mind, which could have rejected it, goes to the newly created memory bank in the main brain. When it gets there the victim recognizes it as a bird because he has an image of the bird in the memory storage of the newly created memory inside the main brain. So an Alzheimer patient has to establish a new information storage just like a little child. The bird he sees is not in the mind of the victim and only knows it as an image and nothing else. He has no other detail of a bird in the memory and sees only the picture and doesn't know if it is a horse or an elephant.

Many scientists believe that Alzheimer is reversible if the required medicine is invented. That is absolutely untrue. No amount of research will come even remotely close to solving Alzheimer disease. Many scientists are encouraged by the fact that victims that had totally lost their minds could be taught simple things and they indeed respond favorably. With adequate therapy and coaching scientist think a victim could be rehabilitated. The only thing they assume so is because a victim with a totally lost mind could be made to understand things like recognizing the bird. But the bird was not recognized by the victim who had lost his mind through his mind but through his newly created memory in the main brain. That is why the victim doesn't know anything about it. He doesn't have the mind to wonder what it is so he just stares at the image of the bird.

The person who lost his mind is irredeemable. He is lost forever. However, that is not the end of the progression of the disease. When the mind finds out that it is being circumvented and never consulted again to retrieve detailed information from its memory bank it desires to reclaim it and to prevent the brain memory signals from going to the main brain memory.

The mind takes drastic measures. It discerns the details in its memory bank and finds a solution. It realizes that the memory bank in it contain details of the various organs of the body. It should be known that the mind has full control over the functions and protects all the main organs of the body. The liver, heart, kidney, lungs, pancreas, esophagus, stomach, intestines, prostate, ovaries, eyes, ears, nostrils and so on and so forth. However, the nerves, muscles, bones, the entire brain, ligaments, nails, the skin, the fingers, arms, legs, feet, blood and a few others are controlled exclusively by the main brain. The main brain has no knowledge of the main organs of the body which is exclusively and only under the directives of the mind.

When the mind discovers the details of the main organs in its memory bank it is delighted. It never knew that it had such full control over the main organs. That is why man doesn't have any knowledge of his main organs and doesn't even discern their exact locations on the body. Those that know the locations of their organs do so by feeling them. Those organs he can not touch he has no clue where they are located in his body. For example, no one could locate the exact locations of his kidneys. If he sees the location on pictures he could only guess. That is why the mind never knew about the various organs before the man was struck by Alzheimer. Alzheimer patients discern and could locate the exact locations of their main organs and could draw around them on the skin to the nearest millimeter. Those who had not lost their minds completely at the time the mind discovered the main organs could verify this fact.

So the mind makes this important discovery and knows exactly what to do. The mind discerns the messages of distress calls coming out of the main organs and compartmentalizes them. The messages from the lungs on one compartment, messages from the heart on the second compartment, the liver, kidney etc. on their separate compartments. When an idea of the availability of deadly toxins in the blood comes from the kidneys, the mind sends that message to the memory brain. The memory brain searches and searches throughout its memory and ultimately forms a good picture of the problem and sends it back to the mind. The mind receives the brain memory message and deletes it. The mind doesn't search its memory bank storage to discern the details and plan of action required to combat the deadly toxin that is in the blood. As a result, the kidneys remain inactive and wait for instructions from the mind and that never comes to them ever. Once the mind desires to delete a particular message from a particular source it rejects any idea that comes from that source to eternity.

So the blood remains infected with deadly toxins. Similar action is taken against the other main organs also one by one. Ultimately the mind desires to tell all the main organs to deliver their messages to the main brain. The main brain has no direct link to the mind and so that plan never takes place. So the mind searches its memory bank all over again and finds many other messages stored in its bank. The search was not a waste of time. The mind discerns messages mainly regarding the heart. It desires to discern messages from the other organs and knows that the heart is the main reason for them being alive. The mind knows the role of the heart very accurately. So the mind desires to reject all those messages returned from the brain memory emanating from the main organs except that of the heart. The mind realizes that without the heart its role is finished if the man doesn't stay alive. So it protects the heart with vigor. That is why all Alzheimer victims have perfect heartbeats.

Unfortunately, the mind doesn't realize that rejecting compiled messages from the brain memory of ideas initiated in the other main organs could kill the Alzheimer victim. As a result, the other main organs, except the heart, decay because their distress calls are not answered by the mind to deal with what ever deadly circumstances they faced. And that is why Alzheimer victims die of multiple organs failures. No Alzheimer patient ever died of heart problem. Their hearts are perfect and should be donated when they die. No heart could be more perfect than that of an Alzheimer patient.

Alzheimer is a deadly disease that could be prevented easily. No one should have Alzheimer ever. People should live up to and more than a hundred years without losing their minds if they could just make a few necessary efforts. No one knows the benefits of marijuana but it is the simplest way of preventing Alzheimer or delaying its progression. Smoking a joint of marijuana every two to three days is the remedy that would prevent Alzheimer from inflicting a person ever. Those who do they will never have any side effects in their entire life. Marijuana is the safest hallucinogen and does no harm to any part of the body. The misconception of toxins in marijuana is utterly false and has no foundation. No marijuana smoking individual has ever experienced any Alzheimer symptoms how much more to be inflicted by the disease.

When a man smokes marijuana the toxic compound inside it is absorbed by the blood vessels by the small intestines. The toxic marijuana molecule in the blood is very deadly and alarms the liver. Bile flows into the liver to trigger it into action to destroy the toxic marijuana molecules. Bile is the secretion of the gull bladder and directly fuels the food the liver requires to do the fighting. The gull bladder does not do the fighting but produces secretions that kill the toxin molecules. So the liver exerts all its efforts for marijuana toxins are a threat to all organs and tissues of the other parts of the body. Many deadly toxins are not as deadly as that of marijuana and that is why the liver alarms the mind immediately. The presence of a very toxic marijuana molecules is transmitted as an idea to the mind.

The mind sends the idea from the liver to the brain memory for processing. Remember, the mind has no idea what the content of the liver's distress calls are and doesn't have the capacity to decipher or decode it. Only the brain memory could. So the brain memory receives the distress message from the mind and releases its search engine scanners to scour its entire memory storage. Eventually, it produces a picture from the messages of the liver that were sent to it by the mind. The mind receives the complete brain memory conclusions and would understand of the presence of a very toxic substance in the blood.

The mind scours its entire memory bank and finds all the details about the toxic substance. The mind immediately transmits it instructions to the liver and orders it to disintegrate the toxic marijuana molecules to smithereens. The Liver musters its effort and obeys the orders of the mind. It delights in the instruction. The liver produces very deadly fire that disintegrates the marijuana molecules to nearly three hundred components. The liver fire could be compared to volcanoes. The deadly toxicity of the marijuana molecule is utterly destroyed. However, the dispersed particles of the marijuana molecules begin to stick to each other. Eventually, two hundred parts converge together and the other one hundred parts also converge together. Hence, the marijuana molecule is reduced to two components.

The liver sends a safety signal to the mind telling the mind that it had reduced the toxic marijuana substance to two components. The mind again sends this idea to the brain memory and after processing sends it completed findings. The mind receives the brain memory result and understands what they meant. The mind searches for details of the information it received from the liver, via the brain memory, in its memory bank. When it understands every information that is available to it, it immediately sends its instructions to the liver. The mind tells the liver to send the two hundred clusters as food to the cells and discharge the one hundred clusters as waste. And that is how the toxic marijuana molecule is converted to a very vital food for the cells.

The cells in the entire body scramble for the new marijuana cluster for it is very nutritional. However the hallucinogenic effects remain intact in the marijuana cluster and all the cells in the body vibrate very, very fast. They all send distress signals to the mind simultaneously. As usual, the mind cannot decipher and decode the ideas sent by all the cells of the body. Hence, the messages of the entire cells are sent to the brain memory for analysis and to assemble a complete picture of the event. The result of the brain memory is sent back to the mind and it understands what the distress was all about. However, the mind wants to know all the details pertaining to the problem and hence accesses its memory bank.

Ultimately the mind finds a solution and sends it back to all the cells of the entire body immediately. It tells the cells of the main organs to simply calm down and reduce their vibration levels. You should know that the mind receives messages from every cell of the body but deletes all messages except those that emanate from the main organs it is designed to control. The same ideas don't go the main brain. The main brain's actions don't come from the distress signals of the vibrating cells but rather from the decisions taken by the mind to resolve the excessive vibrations of the cells in the entire body.

The instructions of the mind calms the cells in the main organs of the body but the other cells that are controlled by the main brain continue to vibrate excessively. When the calming messages from the mind go to the main organs the same messages to the main brain also. The main brain is knowledgeable of all the secrets of the mind and stores it in its memory. So when the mind sends its calming instructions to the main organs the main brain desires to know why all the cells in the main organs desire to be calmed down. In its search, the main brain finds out that all the cells that it controls also required calming.

So it sends the same signal as the mind and that is how all the cells in the entire body are calmed down. The time it takes to calm down the entire cells of the body is what the smoker of marijuana perceives as a buzz. It really doesn't matter how much marijuana the person smokes. A few puffs cause the buzz. Subsequent inhaling of marijuana smoke do not have the same buzzing as the few inhaled at the beginning. Smoking marijuana does not have to be for a long time. A few puffs could produce the required effect to avert Alzheimer disease.

The effectiveness of marijuana is just in the buzz. Subsequent puffs have no effect on vibration of the cells. You can smoke ten joints, one after another, and will never have a buzz. Marijuana clusters remain in the blood for no more than ten hours. If one smokes marijuana within the ten hours he will never have a buzz. The reason is because the cells of the body don't vibrate excessively even if they continue to consume the marijuana clusters.

They will excessively vibrate again only after the ten hours expires and the smoker will now have another buzz after a few puffs. The reason why he gets another buzz is because the cells never find out that marijuana clusters have been exhausted. They never stop vibrating for ten hours only at a much, much slower rate. Subsequent puffs don't produce buzzes because the cells are calmed down by the mind and main brain during the buzz. The time it takes for the mind and main brain to calm the cells all over the body is the time the buzz lasts. Some buzzes last for only a few seconds while others last for a little over two minutes.

The buzz caused by the marijuana cluster trigger trillions of ideas emanating from the fast vibrating cells of the entire body. The mind is nearly clogged with such information and becomes very active. The mind continues to be active for nearly ten hours even if the cells gradually are calmed down to normalcy. They continue to send distress signals to the mind, from the main organs, even though at a smaller and smaller rate for nearly ten hours. The main organ cells also become active when they suddenly began to vibrate violently. Hence, they expend too much energy and consume much more nutrients.

This why when you smoke a little marijuana you get hungry suddenly. The cells use the food molecules available in their surrounding blood so rapidly that they nearly starve to death if the smoker doesn't eat food at all and within a short period of time. The excessive food craving a marijuana smoker experiences is as result of the reason explained above. Marijuana smokers never become overweight even if they consume more food than an average person. The nutrients of the food they eat fizzle out almost immediately and as fast as they eat them. No marijuana smoker can be overweighed.

The buzz a marijuana smoker experiences is not caused by the over activeness of the mind at all and plays no role in its creation in whatsoever form and shape. The constant flow of ideas from the entire cells of the body for nearly ten hours at a time keeps the mind busy even for ten to twenty days. One buff of marijuana could stay in the blood for a maximum of twenty days before they break down into small particles that become very easy for the digestive system of the cells to process. They disappear at a much greater rate as the days pass go by. The marijuana clusters do not make the buzz and knowing that one wonders why they are required to fight Alzheimer disease. The secret is in the ability of the marijuana clusters to trigger vast ideas by the entire cells of the body and would keep the mind busy. Even though the creation of new ideas would be assumed to last for only ten days that is not so. After ten days the mind doesn't return to its normal rate of receiving ideas. The ideas created by the excessively vibrating cells have rippling effects. For example, the initial single idea of vibration message is just that there is a tremor in the entire body. The mind gets the initial message and goes through the entire mind and brain memory relationship and the mind understands there is a tremor in the entire makeup of the main organs.

The mind doesn't have a solution for tremors of the entire cells of the main organs hence, doesn't know how to solve the problem because the smoker is doing so for the first time in his life. But the cause of the tremor is now registered in the brain memory and the mind memory at nearly the same time. Then the mind sends a signal to all the main organs and asks how much vibration they were experiencing. And the cells decipher the request from the mind and send an idea that detail their vibration level.

Since the mind does not comprehend the cell signals it goes through the same mind and brain memory processing relationship and ultimately finds out the level of vibration. It continues to contact the cells of the main organs, nearly a trillion requests, before it finds out that it is nothing serious and merely asks the cells to calm down. This process continues for as long as marijuana clusters remain in the blood of the smoker. As the cells become calmer and calmer, as time passes, the distress signals sent by the entire cells of the body continue to flow at the rate of a trillionth of a second.

Hence, the mind is aware of every step of the calming stage for the ten hours period until the vibrations of the cells return to their normal rate. Persons with such a state of mind would never be victims of Alzheimer disease. Alzheimer disease is caused by a nearly redundant mind. When the mind has minimal ideas to process it begins to scour its memory storage to keep itself occupied and that is a very dangerous situation that all the time result in Alzheimer disease. When the mind is kept busy by the effects of regular smoking marijuana, the mind would never have the thought and time to scour its memory bank and there is no the slightest chance of that person of ever falling victim to Alzheimer disease. That is why Alzheimer disease is the easiest disease to avoid totally and contain even by those who have the symptom already.

The effect of marijuana on the main brain is exactly the same as its effect on the mind and goes through the same process. The mind processes only the ideas that emanate from the main organs only even though it receives such ideas from the entire cells of the body. The mind deciphered at a much earlier stage of the human development that ideas from the main organs are the most relevant ones responsible for maintaining and preserving the mind and soul. If the bodies dies it is the end of the mind and soul on earth. It realized the importance of the main organs and the mind defends them aggressively to sustain its very existence on earth.

The other parts of the body, even though relevant, are not immediate threat to lifespan of an individual. The mind does not even see the entire brain as relevant and has nothing to do with its functions and ignores distress ideas from the entire brain, as it does with the other "irrelevant" parts of the body. A man could live with half of his brain cells dead. But a man cannot live with half heart and the heart is the main concern of the mind and its functions entirely governed by the mind. The other main organs are prioritized according to their importance and their ability to shorten the lifespan of the individual.

Even though a person can live for many, many years with half brain it cannot do so with half of the main organs. A man has two kidneys and could live for a long time with one kidney. When a person has only half of his brain he functions at half of the capacity of the his entire brain and processes only half of the data it receives and other half data don't have any impact in its existence. However, when a person has only one functioning kidney the amount of impurities in the blood do not do not reduce by half. As a result, one kidney ends up doing the functions of two kidneys and it is just a matter of time before it is exhausted and eventually dies. The lungs are the least priority of the mind among the main organs.

A person can live with one lung for a very, very long time. The function of the lung is simply to deliver oxygen to the hemoglobin and nothing else. But the lung becomes important to the mind because less oxygen in the blood is a direct threat to the heart and other main organs particularly. The main organs are affected more drastically by a short supply of oxygen and more than the other parts of the body. If the brain receives less supply of oxygen it regulates its functions and the main brain distances itself from various parts of the brain that are less important to the effective functioning of the person. The main brain does not die before the other main organs that are exclusively controlled by the mind. Hence, if the blood is starved of oxygen because of a malignant lung the death of such a person is primarily caused by the failure of the heart.

The lungs, however, are the least protected by the mind. Distress ideas from the lung are not answered by the mind for a long time and sometimes deleted. Lungs are the most vulnerable and most exposed to diseases more than the other main organs. The heart rarely gets sick because of the ultimate protection by the mind. Any distress ideas emanating from the heart are addressed immediately and remedies instituted instantly. If distress ideas of the heart were addressed in a similar manner as the lungs man would not live half of the age he now lives.

When the main brain receives the same distress ideas as the mind from only those cells of the parts of the body under its jurisdiction, when a person takes a puff of marijuana, it is totally overwhelmed. The mind could process billions and billions of trillion of ideas simultaneously, however, the main brain has an infinitesimal processing capacity of the mind. As a result, when the main brain receives trillions of distressing ideas of the cells of the parts of the body under its control because of their abnormal violent vibrations, it is disoriented. It is clogged by the messages because every distressed cell sends varied levels of signals and has to be addressed individually.

The brain remains disoriented for a maximum of two minutes before it starts to know the solutions to the excessive vibrations of the cells of the parts of the body it controls. That is the buzz a marijuana smoker experiences. It is just a relapse of brain functions for that period of time. The smoker would have a very enlightened mind and would have a fuzzy perception for lack of information in the brain. While the main organs are stabilized by the mind in a trillionth of a second the cells in other parts of the body controlled by the main brain continue to vibrate excessively for sometime before they are gradually stabilized. The buzz may last for a maximum period of two minutes but the hallucinogenic effects of marijuana will continue in some smokers for many hours. Marijuana smokers never experience any variation in their heartbeat and no damaging effects could occur on the heart or any part of the body. Period.

Alzheimer disease is not only controlled by marijuana smoking even though it is the most effective method to stay free of the disease. High grade of marijuana strands should be used to reap the maximum protection of the hallucinogen product. The stronger the better. However, marijuana should never be taken more than two times in a week. And that is the maximum. Those who smoke marijuana more than twice a week could have adverse effects depending on the frequency. A man that smokes marijuana two or more times in a day ends up as imbecile and utterly stupid. He would become so derailed that when you talk to him about Obama he will respond and discuss about horses. The reason for his derailment is because whenever he inhales a few puffs his mind and main brain go through the entire process explained above and how he gets a buzz.

When the same person smokes another joint before the marijuana clusters dissipate all the cells of the entire body experience many of the original symptoms of a first time puff because of the high concentration of marijuana clusters once again. The distress ideas from the cells of the body are not as many as when the person smoked the first time because he smoked before the ten hours effect of marijuana clusters expired. The cells still know the hallucinogenic effects of marijuana clusters. However, the cells totally forget that they ever consumed marijuana molecules after ten hours.

So when the person smokes a second joint in a day the only distress idea emanating from the now slowed vibrating cells is to get rid of the excess marijuana clusters. The mind listens at the concerns of the main organ cells after going through the now familiar mind and brain memory process and issues an order to deal with the new arrivals of additional marijuana clusters. The mind directs that the newly created marijuana clusters be isolated and be devoured by the antibodies.

The antibodies who knew the marijuana clusters as food to the cells and never destroyed them at the first time quickly understand the directives issued by the mind. Blood is considered very vital to the existence of life and hence is controlled and monitored by the mind and not the main brain of the body. Unregulated blood is as vulnerable and likely to cause speedy death just next to the heart. Hence, blood is the next highest priority of the mind next to the heart. Just as it is important to keep a healthy heart so also is important to keep a clean and pure blood.

The mind has full jurisdiction over the functions of the blood and its content and that is why it could direct the antibodies in the blood. As a result, the antibodies devour the marijuana clusters at a very alarming rate and no cluster remains in the entire blood system. The main brain also receives much less distressing ideas from the cells of the parts of the body it controls and is never disoriented. However, it does nothing because the marijuana clusters were wiped out before it could even comprehend the decoded messages it received from the brain memory using the now familiar process of deciphering distress ideas. As a result, the person never gets a buzz when he smokes the second joint of the day. It is apparent that the second and subsequent joints smoked within a day have no effect on the main brain and the senses. Just a waste of money.

When a man smokes a second or third joint he never ever gets a buzz and no disorientation but only feels a clear mind. The distress ideas from the main organ cells received by the mind that ultimately led to the instruction to destroy the marijuana clusters will lead to an overactive mind. The mind continues to decipher more and more ideas that began with the initial distress of the cells that resulted into a buzz and even gets more ideas after the person smokes the second joint. So the rate of data stored in the storage bank of the mind increases dramatically. Unfortunately ninety nine percent of the stored data is silly information and no relevance to the wisdom or existence of the person. Pure garbage.

The person who smokes two or more joints in a day has an overtly active mind and continues to increase the activity of the mind as he smokes more and more joints. The second day, this same person wakes up in the morning and continues his routine. Remember he did not get a buzz when he smoked his second and third joint but only increased the capacity of the memory storage bank of the mind with silly ideas. When that person puffs a few times the next day he gets a buzz again. Marijuana's hallucinogenic effects are supposed to last for at least four to five days and could ever remain in the blood for twenty days in some people. But when the person smokes his second joint in a day the same process goes as with the first joint he smoked with regard to mind brain memory relationship. However the mind's final instruction is to devour the marijuana clusters in the blood by the antibodies. As a result, when the person smoked the second joint he caused the elimination of all marijuana clusters in his blood. Only the over activeness of his mind is the final outcome.

So the next day, the same person smokes another joint. The process is the same as the steps described above when the person smoked the first joint. On the second day, however, the blood had no marijuana clusters throughout the night and the next the process begins as if the man had never smoked marijuana in his lifetime. The cells have very short span of memory and never recognize the marijuana cluster as the nutrient that caused their violent vibration.

Hence, the same process takes place as the first day with the blood of the smoker ending up free of marijuana clusters when he goes to bed. When he smoked his first marijuana joint the second day he would experience a buzz also. And so would be the effect of marijuana to a person who smokes marijuana two, three or more times in a day. Even if you smoke marijuana once ever day the effect is the same. The first day you get a buzz, the second day you only succeed in wiping out the marijuana clusters from the blood and you don't' get a buzz. The third day you get a buzz.

The mind of a person who smokes two or three times that is in danger. Every time the person smokes a joint that person increases the activeness of the mind with silly ideas to decipher. If the person smokes three joints a day then the rate of silly ideas fed into the mind are increased three fold. If the person continues to smoke multiple times in a day for a minimum of two years the amount of processed silly ideas and stored in the mind's storage bank in mind bugling. The Person becomes clogged with silly ideas in his mind and wise ideas become few in number. The mind finds it difficult to access wise ideas from its memory bank and the person will have a sickly derailed mind. When you say something amusing to him he responds to you with a foul words.

Silly ideas are nothing but fast moving messages we see and feel and experience everyday. For example, you see a Jesus monument that all of us see all the time. We don't have any opinion other than knowing that it is merely a statue. When you smoke marijuana many times in day the statue remains the same in the memory storage bank of the mind but instead of just leaving it as the statue that it is, the mind of the multiple smoker would want to know more about the statues of Jesus. Subsequent information are all silly ideas.

The multiple marijuana smoker would want to continue to know more and more information about the statue and will ultimately believe that it is a moving being. Additional information that will be gathered from a single vital information are what will become silly ideas. Normal people will see a statue once and know precisely what it is and ignore it and that is the only image that will be registered in the storage memory bank of the mind. The mind of a multiple smoker of marijuana will have trillion of mind memory storage on that same and simple statue of Jesus. All garbage.

If a person smokes marijuana multiple times in a day, for a minimum of two years, continuously, he should never stop taking it for life if not at the same frequency. If the regular man stops abruptly the mind is shocked and delights scouring its memory storage bank when it is not fed with constant ideas when the person was a regular marijuana smoker. As a result, the mind finds alarming number of silly ideas in its memory bank. Hence, it begins to discard any idea that it feels is silly and does not store it in its memory. That is how Alzheimer symptoms begin. As a result, a person who smoked marijuana for multiple times a day is twenty times more vulnerable to die of Alzheimer disease when he stops smoking the weed.

Many Jamaicans smoke marijuana regularly. Young Jamaicans are vibrant and energetic while smoking the weed. But when they grow older they quit smoking weed and retire peacefully. Those vibrant and energetic Jamaicans end up being docile, forgetful and slow in all their activities. Their mind is blurry and many of them end up looking like zombies. Such people should begin to smoke weed once again before they fall victim to Alzheimer. They don't know that they are in the first stages of Alzheimer. The only reason why they don't feel the impact of Alzheimer that is in them quickly is because their mind memory storage is full of silly ideas that the mind will take many, many years rejecting all the silly messages of the statue of Jesus before it finally rejects the image of the real statue entirely and the person loses his mind.

Marijuana is deadly but the best medicine for the mind if smoked wisely. Students taking difficult courses would be advised to smoke marijuana at most once every month. They would increase the capacity of their minds and would absorb and store in their mind memory banks much of the information they hear in lectures. Those who smoke strong grade of marijuana every two weeks are guaranteed of Alzheimer free lifespan and their minds will be active throughout their lives.

Those who would do so would be able to remember things they did at the age five when they are as old as ninety years or more. No children below the age of thirteen should allowed to smoke marijuana even once. The mind of a child below the age of thirteen is already fascinated by daily discovery of his environment and the discerned ideas he perceives that would finally become the core base of the principles of his life. The size of the mind of a child is not evolved fully until the age of thirteen. Any induced silly ideas derived from marijuana hallucinogens would have deadly repercussions in the ultimate core principles that he would establish that he would guided with for life. Such children would delight in fantasies and unfounded perceptions and become delirious.

Now that we know a little bit about marijuana how about other remedies for Alzheimer? Alcohol is the next beneficial ingredient to fight the scourges of Alzheimer. Alcohol molecules are not that complex. They break down very fast. When a person drinks a glass of whisky the same process takes place as a marijuana molecule. The toxic alcohol molecule is broken down to two hundred smaller molecules, just like that of marijuana. However, the tiny alcohol molecules do not form clusters like marijuana when the liver disintegrates them into tiny bits. The tiny molecules of alcohol are then desiring to join one another to form into the original large alcohol molecule they were disintegrated from. But that never happens and hence, two hundred fragmented molecules flow into the blood stream.

Knowing that one wonders how they became disintegrated. What is the cause of their fragmentation? The liver desires to know the original alcohol molecule when it sees it in the blood stream. Then it discovers it is very toxic because the alcohol molecule releases a very dangerous secretion. And that is what causes the alarm for the liver to send an idea of distress to the mind. The mind goes through the same process of the now familiar procedures and comes out with an ultimate solution. The mind tells the liver to obliterate the alcohol. The liver has a very powerful fusion system that produces a deadly laser that disintegrates any enemy perceived by the mind. So the blood system is clogged with tiny particles of alcohol. The problem is that the antibodies don't see the tiny alcohol molecules as foreign particles. They don't see them as such because the antibodies don't perceive them as a threat to any part of the body. And they are not. The tiny molecules go into the cells and become food to them.

When the cells consume them they realize that they are hallucinogenic. They begin to vibrate violently as a result. Hallucinogenic means that the molecules of the alcohol hit primarily the heads of the cells repeatedly just like a derailed person would never know the effects of hitting your head all the time. So the cells vibrate violently because they remain stationary while the alcohol molecules repeatedly bang their heads. You can compare the same way as a fight in a boxing ring. If one of the fighters remains stationary and received continuous blows to the temple, his heads will become disoriented and seizures take effect. The body will react and the result is body shaking. That is all vibration means. All vibrations are caused as a result of the molecules continuously hitting the heads of cells. Knowing that observe animals fighting. Every animal knows and targets the head in order to inflict maximum pain upon its fighting partner. Chickens kill a poisonous snake by pricking the head of a snake. Goats bang each other's heads. Even humans know the head disables a person much faster than hitting him on the legs.

The alcohol molecule doesn't know that it is inflicting pain upon the cells. That is what delights the alcohol molecule the most just like a human delights in fingering his hair. Mere passing a man's fingers through his hair dishes out exciting messages throughout his body. That is all. The alcohol molecules think they are just bouncing balloons on the cells' heads with no intention to destroy them. As a result, the mind returns its final decisions that were derived from the vibrations of the cells back to the cells. Instead of asking them to calm down, as was seen in the vibration caused by marijuana molecules, the mind asks them to make no more vibration mainly because the blows on the head are not as severe as that of marijuana molecules. However, the cells continue to vibrate because they received very hard blows and couldn't stay stationary. They send the same distress signals again to the mind. The mind sends them back the same instructions again and to stop vibrating.

The cells become frustrated. They then make up their minds and desire nothing to do with the main mind that controls all the organs and blood. This takes place only in the kidneys and in no other place. The kidney is the organ that identifies the fragmented molecules of the alcohol toxin. So the kidney takes the law into its own hand and devises its own method of solving the problem. The kidney is extremely delicate. A slight cut on its outer surface will cause profuse bleeding and eventually drain all the blood in the entire body. Blood vessels are even on the surface of the kidney just a thin layer covering them. That is why the kidney is red in color and receive its redness from the oxygen molecules carried by the hemoglobin.

Hemoglobin are dark charcoal in color. However, oxygen atoms emit purely red color and are the red color on the rainbow spectrum. That is why when a person bleeds the surface of the blood spilled is slowly turned into a black mass of crust. It you remove the top layer you will find purely red blood. And when the exposed part is further exposed to the atmosphere it becomes dark again until the entire blood becomes black as charcoal. The reason for the final darkness is because as the oxygen in the hemoglobin's body disperse into the air it turns darker and darker until all the oxygen molecules disappear. That is why blood turns into dark mass after a few minutes of exposure to the outside world because what is left is pure hemoglobin only.

So the kidney which is basically blood vessels only that make up its body devises a crude way of getting rid of the toxins in its body of a wealth of blood vessels. The kidney expands its capillaries and begins enlarge slightly. As the blood vessels expand the blood flows easily. The kidney cold expand and constrict its capillaries and no other organ of the body could do that. It is purely kidney function and that is all it can do and no other thing. No other organ of the body has the capability to do that. So when the kidney expands the blood flows easily in its slightly enlarges capillaries. This always results in dispersing the toxins over a larger area no matter how slightly. The toxins become very familiar with their new environment immediately. They live more comfortably and swim in the blood over a larger area.

Then the kidney contracts the blood vessels at a faster rate than it expanded. And the toxins become disoriented. They desire to get out of the kidney but they could not because the capillaries are so tight. Gradually the kidney loses so much blood because the capillaries are small and normal blood flow is disrupted. The blood is gradually drained out and new blood is now almost at a halt and the kidney turns into grey. The alcohol toxins use the oxygen in the blood but now there is virtually little blood in the kidney capillaries. The cells of the kidney also use the oxygen from the blood. So the effect from the slowing of the blood is the same for the alcohol toxins and the cells of the kidney. The alcohol toxins die faster than the kidney cells because the kidney prepares its cells for such eventuality. Prior to the expanding and constricting of the kidney capillaries it advises its cells to extract and store more oxygen molecules. When the kidney cells have three to four oxygen molecules in their body, the toxin cells would have the normal supply of cells of only a maximum of two oxygen molecules.

The alcohol toxins use their oxygen supply and cease to exist after that for they would have no more supply of oxygen because of the lack of blood flow. So the cells of the kidney outlast those of the alcohol toxins. And that is how the kidney gets rid of toxins in general. But that is not the only way. It has other two methods of wiping out impurities in the blood.

One other method is very simple and straightforward. That is the primary method of eliminating poisonous toxic molecules. When the molecules arrive at the kidney and fill the capillaries the kidney instructs all its cells to never desire them. This instruction comes from the mind after the kidney sends distress idea to the mind when it observes those toxins. The mind processes the idea and sends instructions to the kidney after consulting with the brain memory and receiving the final processed data. By now you are familiar with the mind and brain memory relationship and how the mind arrives at its final action plan. Hence, no need of repeating it frequently.

In this method the cells of the kidney repel the poisonous alcohol toxin and isolate them. Hence, the toxin molecules gather into loosely attached clusters. The clusters are pushed by the various kidney cells further and further until they find themselves in the center of the kidney. That is where the deadly secrets of the kidney lie. The interior part of the kidney is comparable to a nuclear reactor. When the toxins finally converge in the center of the kidney gamma rays are produced by the membranes that make up the center of the kidney. As a result powerful beams are produced which obliterate the alcohol toxins to almost atomic parts. That is the safest way of eliminating poisonous toxins by the kidney. This is the normal procedure and primary function of the kidney and this could only occur when the kidney is exclusively instructed to do so by the mind. The other two methods are devised by the kidney, unilaterally, when the mind ignores the distress calls it sends to the mind. The parts of the kidney that produce the gamma rays are called nephrons by today's scientists. The obliterated parts are sieved of their hemoglobin contents and urinated.

The third method is deadly and is the last resort taken by the kidney and sometimes with a very adverse effect. When the kidney capillaries are filled with toxic poisons the kidney sends a very, very loud distress to the mind. The mind receives the distress idea and sends it to the brain memory and receives back a final and compiled message from the memory brain. The final instruction from the mind is not favorable to the kidney.

The mind tells the kidney to resort to its nephrons and obliterate the toxins with laser beams. However, the kidney realizes that the toxins are too large and its cells could not repel them enough to form clusters to be pushed to the center of the kidney where the nephrons are located. The kidney sends back it inability to defend itself to the mind as the usual distress idea. But the mind doesn't have any other option of dealing with such a dire circumstance. As far as the mind is concerned and the data in its memory storage bank that is the only function of the kidney. To obliterate foreign objects using laser beams. Hence, the kidney is forced to devise its own way of dealing with the situation without the knowledge and instruction from the mind.

This third method is very important because it is mainly used to discard alcohol toxins. When the toxins infest the blood in the kidney it seals all the entrances of the capillaries right at the main entrance. As a result and within a short period of time the entire blood in the kidney flows out of the kidney and no new blood is pumped in because all the entrances have been constricted not to allow any blood flow into the kidney. The kidney does this blindly and doesn't know of the evil repercussions that could result. The cells in the kidney were not told to prepare for the action the kidney decided to take neither were the alcohol toxins. As time passes, the alcohol toxins begin to die because of lack of oxygen. In a similar way the cells of the kidney also die moments after the alcohol toxins are exterminated. In essence, the kidney commits suicide and that is why we have kidney failures.

Alcohol molecules are not deadly and don't damage any of the main organs and other parts of the body in whatever way or passion. Too much alcohol consumption affects primarily only the kidneys and pancreas. It does very little damage to the other parts of the body including the brain memory and the main brain. The kidney did not have to resort into the third method and could have used the first method to get rid and successfully eliminated the alcohol toxins. The main reason why it resorted to the third method is because alcohol molecules are too large.

Alcohol molecules are so large that you could see them with modern microscopes to verify. The first method would have been difficult because when the blood vessels constricted the alcohol toxins would have dived away and would have discerned the real intentions of the kidney when the capillaries began to constrict. Hence, they would have flowed out of the kidney when the blood dissipated gradually and none of them would have died. The kidney devised the third method because it tried to use the first method repeatedly and unsuccessfully. The kidney has no memory organ. It just relies on the messages of the mind. The methods were devised not by memory experience but by sheer experimentation. Knowing that, alcoholics have ten times of kidney failures than regular people.

The nephrons are deadly ligaments looking membranes. They don't have cells and don't have memory. They make gamma rays by splitting nitrogen molecules. The temperature of a nephron could reach up to 2000 centigrade if measured. However, such temperatures are too minute to hurt any part of the kidney. Many nephrons emit gamma rays when the nitrogen molecule is destroyed to its atomic component just like uranium in a nuclear explosion. Many particles that remain in the kidney after the destruction of the alcohol toxins are deadly. Their makeup is ultimate disease to the interior of the kidney. That is why those who die of kidney failures have dark plucks in the interior part of the kidney. Pluck from kidneys is very, very, very toxic and should never be touched by bare hands when extracting it for autopsy. It radiates even after death no matter how small.

Now we go back to the importance of drinking alcohol. The above explanation is merely what happens when alcohol is being disposed off from the blood vessels. Alcohol is very relevant to the functioning of a normal body. Man does not live without the help of alcohol ingredient. Everyone manufactures alcohol from many, many sources. Even animals live on alcohol. Even if you don't drink any alcohol the mind manufactured alcohol from the different types of food you consume. No one knows it but every living being is daily tipsy even if he doesn't stagger. Manmade alcohol is a deadly poison because of its composition. Knowing that you should learn to regulate the amount you consume for the maximum benefit of alcohol.

When man consumes alcohol it is for a very special reason. The mind doesn't reject alcohol in the blood stream that is manufactured by the body and from the food that is derived from one's diet. However, it vigorously discerns the toxicity of manmade alcohol and acts appropriately to discard its toxic component. Manmade alcohol is made purely to do the same functions as that made by the body. That is all. When you drink alcohol your mind delights because newly created ideas are received by it. It becomes hyperactive and that it is healthy and that prevents the chances of degenerating and becoming Alzheimer victim. Alcohol consumption works in the exact fashion as marijuana smoking. Everything derived from marijuana benefits are also true with the alcohol consumer.

When a man/woman drinks a few shots of whiskey in day he doesn't know that he is messaging any parts of his body. He drinks it for pleasure and truly makes him tipsy or drunk and delights in his state of mind. Many alcoholics have no memory of their state of mind of events that took place a day before. Their state of mind is renewed every day. So is a fact with marijuana smokers. If they could remember what happened to them in the previous day none of them would desire to drink again. It is not because they develop memory relapse but their situation is so insignificant and are not traumatized by it. Hence, it becomes so minute and is registered as a highly insignificant and extremely tiny silly idea in the storage bank of the mind. An alcoholic could remember some part of what he experienced the previous day and delights in it but doesn't recollect everything that took place when he was drunk.

Alcohol molecules' destiny is always the pancreas. The pancreas doesn't produce insulin as falsely believed and has only and only one specific function in the entire structure of the body. Pancreas breaks down alcohol particles in its useful components. Pancreas produces an enzyme that discerns the complexity of alcohol. This enzyme is responsible for the final result of the molecules that return to the blood steam. If there was no pancreas the many alcohol large molecules will desire other organs and in the process destroy many of them. Luckily, no alcohol molecule is absorbed by the cells of the other parts of the body and the large alcohol molecules would not be able to penetrate other cells in the rest of the body. Only the pancreas has porous cells that are wide enough to provide entrance to the large alcohol molecules.

Many of the alcohol molecules disintegrate upon contact with the enzyme cells of the pancreas. These molecules are still too large to access other cells in the rest of the body. As a result, the pancreas discerns this fact and releases additional enzymes. The additional pancreas enzymes further break the alcohol molecules even to its smaller component. Remember that the pancreas enzymes don't have the ability to change the molecular structure of the alcohol particle. They only reduce them to a tolerable size to be absorbed by other cells of the body and particularly the heart cells. The heart cells are not so tiny and are fairly large compared to the other organs of the body.

That is why the kidney cells don't absorb alcohol molecules even at the least sizes because the kidney cells are so tiny and their pores are also too small for the alcohol molecules to access them and use them as food. Or else, the kidney cells wouldn't send distress ideas to the mind at the presence of alcohol molecules in its blood capillaries. The reason why it does so is because even the now reduced molecule of the alcohol could easily block smaller capillaries and eventually do so if they formed a small cluster of alcohol molecules. Alcohol molecules are a direct threat to the very existence of the kidney and nothing else.

Alcohol consumption does not access any of the other parts of the body except the heart. Alcohol does not destroy and single cell in the body no matter how much it is consumed. It only damages the heart and the kidneys stupidly imagine that the alcohol molecules would form clusters that would kill it permanently even though it never ever happened. Kidney failures would never occur if the kidney was a wise organ. That is why the mind does not find any fault in the presence of alcohol molecules in the capillaries of the kidney and does not share its derailed and imaginary perception of a danger that does not exist.

The now reduced alcohol molecules flow into the blood stream after exiting the pancreas. They search the various parts of the body to enter the cells to expend the powerful energy that is in them. They search and search. Some of the molecules desire to exit from the body and they succeed. That is why drunkards have alcoholic breath when they consume so much alcohol. Whether a person consumes whisky, vodka, gin etc. the alcoholic breath of a drunkard has the same odor across races. The alcoholic molecules eventually find the appropriate candidate that accept them as they are. Alcohol molecules know that they are very relevant to the body to provide good and powerful energy. They enter the heart and are immediately devoured by the heart cells. The heart cells send distress calls to the mind because the alcohol molecules they consumed provided them with too much energy than they could handle.

Remember that marijuana molecules had hallucinogenic effects and knocked the heads of the cells that resulted in the distress call to the mind which caused excessive vibrations in the affected cells. Alcohol molecules don't have hallucinogens in them but are loaded with energy because of the carbon that exists in them. The carbon is released from the molecular structure of the alcohol molecule when bombarded with laser like accuracy produced by the internal combustion of the heart cells.

Assume alcohol molecules and compare them with charcoal and marijuana particles with a helium balloon. Also remember that alcohol molecules don't make the heart cells vibrate excessively which could kill them if it is not arrested quickly. The heart cells just have too much energy in them which does not kill them. If a man consumes too much sugar would have excessive energy, for reason that would be explained later, and does not feel his life threatened. Exercising would be the most appropriate thing to do to expend such energy but most people don't even know that they had such a large energy in them.

That is why autistic kids show a more volatile behavior when given too much sugary food or are on constant carbohydrate diet. Such kids can't handle the change of energy that is present in them and they move different parts of their bodies to get rid of the excess energy. Constant energy is very important to the body cells but when it exceeds to a certain degree it becomes a discomfort. Autistic kinds feel as if their bodies are coming out of their skins. And that is how the heart cells feel. They feel as if their outer membrane would disintegrate which utter falsehood. Like the kidney cells the heart cells also have a weird perception.

The distress calls sent by the heart cells to the mind are so silly that the mind ignores it. When the mind receives the finalized picture of the of the distress ideas from the brain memory, after the now very familiar procedures, the mind scours its memory storage bank and easily finds the solution it had delivered to the heart cells earlier. The mind sends its finalized decisions to the erroneously distressed heart cells and instructs them to discard their silly ideas appropriately. Hence, the heart cells no more send distress ideas to the mind regarding that particular issue of high energy content in the cells' body and instead desires to solve the problems themselves.

Remember, the kidney also did the same thing when its distress calls were ignored by the mind and sometimes resulted in the demise of the kidney. The heart cells, however, don't have the capacity, as the kidney, to institute any changes to its ability. As a result, the only thing the heart cells could do is to remain as stationary as possible. But that has a repercussions. When the cell becomes very energetic as it consumes more and more of the high energy alcoholic molecules, it leaps upward and then returns to its original position. The energy released by the leap is enough to reduce the excess energy in the heart cells. The heart cells realize that it is the best way of preserving their structures and the erroneously perceived possible ultimate death. As a result, they all leap and return to their original positions at varied intervals and depending on the level of energy content of each cell. That is why all drunkards have heart murmuring. Heart murmuring is no symptom to a possible heart attack that could occur and is certainly not as a blockage in the capillaries, veins or arteries.

Now that you know the good aspects of alcohol does it have bad results when consumed? The answer is, yes. When the pancreas reduces the alcohol molecule to that the heart could admit into its cells, it becomes jealous and wants to also consume it and derive its unique quality. Hence, it continues to reduce in size a tiny portion of the alcohol molecules and quickly admits it into its cells once it is appropriate. The pores of the pancreas cells are just a little tighter than that of the heart. That is why the heart and the pancreas look alike to the naked eye.

The alcohol molecules have the same effects as that of the heart cells. The pancreas cells experience the same false perceptions and send distress ideas to the mind as the distressed cells. However, the pancreas is the least important organ next to the spleen. The pancreas was once important when man ate raw meat which contained deadly toxins. But as man learnt to cook meat and destroy all the toxins found in raw meat, the role of the pancreas became more and more insignificant. Hence, the pancreas is only good for the reduction of alcohol molecules only. That is all. Even then, over ninety percent of the breakdown of alcohol molecules are done by the liver.

The pancreas doesn't want to be wasted away so it desired to ascertain its importance. It discovered that the heart required alcohol molecules of a certain size and it produced only that. The alcohol molecules blasted by the nuclear fusions of the liver are so tiny and ended up being consumed by all the cells of the body even the brain cells. However, the heart didn't derive much energy from the tiny alcohol molecules produced by the liver and instead selectively chose those larger alcohol molecules produced by the pancreas. Hence, the pancreas is a worthless organ and man could do without it even if it dies permanently. A dead pancreas could stay in a person's body even for a hundred years because there is no oxygen outside the cells inside the body which is clearly responsible for the rotting and disintegration of body tissues.

The reason why the pancreas cells send a louder distress ideas to the mind is because the outer coverings of its cells are very thin and delicate. Before the distress ideas are even sent to the mind some pancreas cells begin to burst and die permanently. The mind doesn't value the pancreas at all, at all. The mind feels it is the waste of its time to respond to the pancreas' volatile calls for help. Hence, the pancreas eventually ends up as a dead organ. All drunkards have dead pancreases. But because it is a worthless body organ its death has no repercussions whatsoever on the health and wellbeing of the drunkard. And that is the only adverse result of consuming alcohol, regardless of how much and how little.

There are two types of heart murmuring. The above stated fact is one, however, there is another one. Man named it heart palpitation to distinguish the difference. The causes and eventual results may not be the same but both heart murmuring and heart palpitations are the heart's own way of tackling its problems that would be ignored by the mind. In both cases the heart responds by the cells of the heart jumping up and down to expend their extra energy. Heart palpitation is caused by blocked capillaries, veins or arteries. Pluck is formed on the inner surface of the blood carrying tubes of the heart. There are deadly repercussions when the fat deposit sticks on the surface of the blood carrying tubes of the heart.

When that happens the cells in the heart's capillaries send distress signals to the mind. The mind receives the distress idea and sends it to the brain memory. The brain memory comes up with a complete picture of the idea sent by the mind and returns it to the mind. The mind deciphers the distress idea of the heart cells and gives its final instructions. However, the mind doesn't see the pluck deposit as a threat to the whole heart and hence ignores it. After all, there are just a few capillaries, compared to the giant heart, that were in distress for the pluck initially reduces blood flow to a very small segment of the entire heart cells.

The response of the heart to heart palpitation is the same as when it responded to excessive energy created by alcohol molecules. The heart is deadly silly but the kidney is much wiser. As a result, the cells begin to jump up and down and heart murmuring occurs. But this time the murmuring is confined to the only part of the affected cells where there is a pluck deposit and a reduced blood flow to the cells. As a result, the person feels palpitation in a particular part of his heart where the murmuring occurs. The other parts of the heart don't murmur because they don't see the essence of such an action. Remember that in the murmur of the heart that resulted from alcohol molecules induced problem, the entire heart was affected because the alcohol molecules were everywhere in the heart. Therefore, the drunkard feels as if the entire heart was going to die.

Unfortunately, the person experiencing heart palpitation does not hold the same concern as a drunkard with merely heart murmuring. The same mind which ignored the distressed cells that were nearly starving from nutrients and oxygen scrambles to rectify and rescue the dying cells around the pluck filled capillaries that resulted in the palpitation of the affected areas of the heart. It goes through the process of mind and brain memory relationship and issues its solution and passes its final decision. To pass more blood into the nearly starving cells of the affected it increases the heartbeat gradually until enough blood flows into the affected cells through the nearly blocked capillaries. Therefore the person with the heart palpitation will have a new rate of heartbeat and all the time higher than his previous rate. The pluck gradually deposit more and more and the heart increases the rate of the heartbeat to maintain adequate blood flow to the affected cells. That is how high blood pressure occurs. The only way the person could reduce his excessive heart rate is through dieting. No other way. The solution would be explained much later in this writing.

The reason why the person develops plucks in his capillaries, veins and arteries is because fats are dirt in the blood and do not cause any harm to the entire body and hence, the antibodies don't destroy them. Antibodies know the production of every organ in the body and don't destroy the secretions produced by the various organs. Fats are not produced by any of the organs of the body, however, some of the fats are deposited on the liver, the first time fats enter the body, and the liver doesn't destroy them because they don't any dander to any part of the body. As a result, the antibodies know that if the liver doesn't see the danger it is none of their business to destroy them. That is why fats deposit everywhere they desire and no part of the entire body sends a distressing signals to the mind.

Fats are nothing but proteins. No one should eat protein for it is a threat to the existence of man. Proteins produce only fats and have no relevance to the body functions whatsoever. Fats were made to be deposited under the skin to protect people in extremely cold environments and preserve their lives. Other than that proteins are a death sentence to those who live outside a cold environment. No human being should eat a molecule of protein for there is no benefit and only threat to every part of the body. Organs mainly affected by fat deposits converted from protein are the same way affected as the damages inflicted upon the heart. They all send signals to the mind when their capillaries are nearly blocked and the mind ignores them as the initial distress cry of the affected parts of the heart. The mind responds immediately a palpitation occurs in a small segment of the heart and dictated instant remedy because it values the heart more than any other organ of the body. The mind knows that a dead heart ends its existence on earth and thus the ultimate concern it places in its priorities.

Unfortunately, the mind doesn't have equal concerns, and does not act as fast, when distress calls are sent by the other organs when they experience blockage of capillaries on some parts of their bodies. The kidney is the only one that devises a unique way of resolving such a problem and desires nothing to do with the method used by the other organs. The kidney doesn't have and deposits in its capillaries because it senses the harmful and evilness of fat deposit. Hence, it destroys it as soon as a single protein molecule enters into capillaries. It does so by the second method explained above.

The other organs desire to copy the heart and therefore murmur and cause palpitations in their affected parts of their bodies to delete the fat deposits in their affected capillaries. Unfortunately the mind does not respond, as it did with the heart, when palpitations occur, even though the mind receives loud cries by the near starving segments of a particular body organ. The result is the eventual death of the areas inhabited by such blood starved cells and hence, the slow death of the affected organ.

The deadly effect of protein molecules maim and destroy the pancreas especially. If the pancreas is not destroyed by the energy molecules of alcohol, it ends up dead by the deposit of protein caused fat when it easily blocks its capillaries, veins and arteries. Protein fats cling on the inner surface of the blood carrying tubes because the cells of the pancreas are larger than all the other organs, next to those of the heart. However, the outer coverings of the pancreas cells are very delicate and fragile. They blow up when the heartbeat is increased by the dictates of the mind delegated to the heart. The heart is the most affected by protein induced fat because it has the largest size of cells which become very convenient for protein fats to cling on and form large clusters.

Go to Nigeria and visit a poor village in Yoruba land. Observe all the young children from the age of ten for that is when muscles reach their highest development. Such children have well defined muscles as if they if they do muscle building exercises as do body builders. Nothing of that kind. Yorubas mainly eat cassava, and in a poor village, yam is eaten but is a luxury food. Yam and cassava are primarily the richest carbohydrate sources and consuming them only is the best thing you can do for your body. Yorubas don't eat vegetables and if they do only consume okra. If it wasn't for malaria, which is very predominant in Yoruba land, Yorubas would live to be more than a hundred years with their body strength nearly intact. Never ever has a single Yoruba died of heart attack in a poor villages. It could never happen and it is an absolute impossibility.

Yorubas die in large numbers and become extremely vulnerable when they change their diet and adopt western lifestyle. A Yoruba in a western country is five times more likely to die of heart attack than an European. The first protein molecule he consumes become a pluck in the inner tube of the heart vessels instantly far more frequently than an European.

Many African tribes don't eat vegetables at all. That is why they don't suffer heart attacks in totality. Heart attack is unheard of in many African countries. Those with small numbers of heart attacks live in the cities and consume European garbage diets. The muscle lasts for fifty years as long as is it fed with adequate carbohydrate to sustain energy level for the relevant combustion and to safely exist in the body. After fifty years the Yoruba man would not need much carbohydrate to sustain muscle strength and his muscles gradually wean and would end up with smooth flesh of the body. However, if that child of ten in a poor Yoruba village is transferred to different location and does not get carbohydrate rich food, he will lose his well defined muscles beginning from three days to a maximum of three months. Within just that short period of time a Yoruba body will look just like that of the body of the child in his new environment. That is why Yorubas living in the western world don't have a single muscle on their bodies.

It takes three months to lose your muscles and takes three months to build or rebuild them. Carbohydrate is all you need to eat to live a long life. Those who don't consume enough carbohydrates are deformed people, with no muscles and do not attain the statues of a normal man. Chinese are a typical example. They eat mainly rice which is not rich in carbohydrate and are midgets. Let the Chinese feed exclusively on carbohydrate food and they will grow muscles and reach the height of a regular man immediately. The reason why Chinese are short and weakling has nothing to do with the structures of their genes. Even if genes have a role to play, Chinese could attain six-feet height within just a single generation.

Mainly African people discerning this message will understand the role of diet in their livelihood. Overwhelming majority of humanity will discard it as utter falsehood. That s not a surprise. No one will comprehend it like Africans, especially West Africans. Many will discern it but will not apply it. That is their damn business. The writer's role is to expose the truth and the rest remains in the hands of the readers of this relevant messages.

Protein is deposited in different ways. Animal fat desires to stick on muscles and is deadly People who eat animal fat are the ones that suffer heart attack. The heart is made up of numerous muscles and its blood vessels are venerable to plucks formation on the inner part of the tubes carrying blood. The reason why the deadly animal fat sticks primarily on the heart muscles is because many of the heart blood vessels don't have cells covering its surface like the rest of the organs. The interior part of the heart's blood vessels are daily hungry for nutrients because of the energy the heart expends to do its functions. Coverings on the inner surface of the blood vessels would delay the consumption of food and oxygen and hence, the mind discarded it in its entirety. Food and oxygen enter the parts of the heart at much faster rate than any of the other body parts.

Many fat molecules stick on the interior surface of he blood vessels because of the nearly rough surface of the blood vessels. It is easier to walk on a rough surface and very difficult to do so on a very smooth surface. That is why the heart accumulates pluck on its rough surface of its blood vessels. Humans don't have memory of the heart rate but the mind discerns every heartbeat of the organ. Knowing that you should know that your heart is the safest organ of your body if you don't destroy it yourself. Knowing that, again, you should make effort to protect it because it is the main engine of your life. Knowing that, again, and again, you should desire heavy messages to know the functions and learn to respect it.

Plucks are not dangerous if they don't stick on the blood vessels of the heart. Many fat molecules are not a danger to any of the body parts. Nothing in fat is used in the body and consuming it is utterly silly and stupid. Fat molecules do not serve as fuel to the cells as assumed by researchers and is an absolute falsehood. Dirty fats is what you will end up with at the end of the day. Fats are never digested by any cells of the entire body and serve no purpose to consume them. Proteins are the only source of fat on the body apart from consuming animal fat directly. Sane people should never, ever, eat a single molecule of protein.

Many vegetables have protein in them. Knowing that, you should discern and avoid vegetables that are rich in protein. Protein fats are not entirely responsible for the pluck on the heart even if they sometimes do. But rarely. Vegetable fats prefer the abdomen mainly. People with blotted bellies are abusers of vegetables and consume them in great quantity. No one who doesn't consume vegetables will ever grow fat until he dies. Obese people are just people who enjoy eating vegetables and their metabolism is dead before they are twenty years old. The reason why their metabolism dies is because they no longer produce the required enzymes to breakdown food particles adequately. Food ingested through their mouth is likely to be excreted in almost its raw form.

Obese people don't eat more food than normal people. They don't know how to fight obesity and delight in eating more and more vegetables to do away with their fat. Unfortunately, the more they eat vegetables the more they grow fatter. If you want to live a happy and have a thin body desire no more to eat vegetables. If a fat person doesn't consume vegetables for sometime, the accumulated fats in his body are delirious. They don't know why there is no covering for the outer part of the fatty surface of the body. The surface doesn't desire to remain exposed for a long time. It delights when another layer covers it. So when there is no more fat produced because the fat person stopped consuming vegetables, the topmost layer on the outer surface of the fatty area begins to disintegrate. Fat molecules begin to desert the outer surface of the fat area and enter the blood stream. These fat molecules don't look like the original fat molecules because they have become larger while attached to another molecule on the fat area from where they extracted themselves.

Hence, an obese person could lose large part of his fat concentration by just not consuming vegetables. As simple as that. Knowing that, no human being should become fat enough to loose mobility. Having simple diet will save your life until you die. All you have to do is eat exclusively carbohydrate rich food and avoid consuming protein entirely. If you don't die from the failure of other organs you are guaranteed not to have heart attack for the entire period of your existence on earth. Those with palpitations I their hearts should know that they could reverse their deadly situations instantaneously. Just stop eating vegetables and will have no plucks on the heart's blood vessels which are the cause for heart attacks. If all humanity avoid eating vegetables all heart surgeons will only be hired as bricklayers for they know nothing other than operating and dissecting the hearts of unfortunate victims.

Knowing that you should only carbohydrates rich foods only what would happen if you suddenly stop eating vegetables. Nearly nothing for you will only become a much healthier person. However, minerals are very, very, very essential to the functions of your body. But that could be compensated easily. Take vitamin and mineral supplements. They nourish your body system just as those vitamins and minerals that are derived exclusively from vegetables. Otherwise not eating vegetables for life will have no impact in your development, growth, functioning and your daily living. Avoid vegetables like hell. Lettuce is very bad for your system and has carcinogen that is deadly to the cells. Cabbage is another garbage. Those who eat a little cabbage become sick in their eyes.

Carrots have no relevance to the body and carotene does not aid the eye whatsoever. Beetroot, very well cooked, is a healthy vegetable and contains no protein and is rich in carbohydrate. Spinach rich in iron but very rich in protein also and you should avoid it. You don't need constant iron supply in your body. No hemoglobin can exist with iron atom. But every hemoglobin that has iron keeps it for as long as twenty years. Hemoglobin don't die and could live for as long as that period of time, twenty years. The mind regulates the production and functions of hemoglobin.

The hemoglobin is produced in the bone marrow of a person. As a result, hardly any new hemoglobin are manufactured by the bone marrow for a long time. And when it does, there is always a reserve in the bone marrow tissues to supply the new ones. So, you should take one or two iron capsules every six months and that is more than sufficient. Taking more iron supplement will not harm you and will be discharged as a waste matter in your urine and will only see greenish a color in the discharge because iron atoms combine easily with protein fat and form a new compound resulting in the change of appearance in your fat. Bile in the liver has dark green appearance because of the high concentration of iron in its secretion. Bile molecules are ninety nine percent iron compounds and one percent fat.

Having said that, how would one know what vegetables to eat and what to avoid. The simplest way is to perform an easy experiment in order to categorize vegetables according to their importance to the body. A goat has the most sensitive membrane in its nose and could identify protein content in a vegetable with utmost accuracy. Goats know the toxicity of protein to their body and do not eat any plant that contains any protein molecule. Starve a goat for two days and then line up various vegetables fairly apart and then release the goat to feed on them. The goat will smell each vegetable and every time it sniffs when it smells a particular vegetable you should know that, that vegetable contains a toxic level of protein. The vegetable the goat devours with ease that is what you should recommend for human consumption. Only African goats should be used for such an experiment, preferably a Barka, Eritrea, bred one.

Humans never found a solution to heart attack until now. No one should die of such an affliction that could be avoided in its entirety. Blood thinners are sickest thing you could prescribe to a person with a palpitating heart. All what blood thinners do is breakdown the protein fat molecules in the blood to two parts or three. Nothing more than that. However, such a victim is always placed on strictly vegetable diet. That certainly creates more protein fat molecules in the victims blood. A doctor prescribes a higher dosage of the blood thinner when he realizes that his blood was congested with high cholesterol. Cholesterol in nothing but protein fat.

When a person eventual swallows a 50 milligrams capsule of Lipetor, the drug breaks the protein fat molecule in the blood to its tenth part. Hence, the tiny protein fat molecule could easily cling on the inner surfaces of every blood vessel in the entire body. Instead for the kidney to discharge the protein fat molecule through urine, all the deadly protein fat molecules become part of the blood vessels. They build plucks immediately when one protein fat molecules sticks on the inner surface of the blood surface and subsequent molecules attach to it. Protein fat molecules easily reattach themselves with one another daily and by the end of the day a tiny dot the size of the sharp end of a pin is formed. As more and more Protein fat molecules, cholesterol, mount one upon another they form a thick membrane that restricts the fine flow of the blood in the capillaries first and then the veins and arteries.

The large protein fat molecules are considered to be considered deadly and the victim is deemed to have high cholesterol. Those broken down and tiny protein fat molecules are considered, achieved with high dosage of Lipetor, are said to be not only very good molecules but also very relevant to the functioning of the blood in the body. That assumption can never be further from the truth. The truth that is the original protein fat molecules which are larger in size than the medication reduced size of the cholesterol, do not have the mind to stick to any parts of the body except the heart which has a relatively rougher inner surface of blood vessels. Hence, more than two third of the protein fat molecules roam inside the blood vessels indefinitely. When they get tired of roaming they only one part of the body that they could stick to. And that is the gut in the belly of a person

The protein fat molecules easily find the belly convenient simply because the linings of the outer surface of the intestines are made up of merely big pores to allow easy flow of nutrients to the blood system. The large protein fat molecules easily flow through these big pores and delight in doing so. From there they delight in doing something else too. They find very far away a set of molecules that combine with them. These molecules derived from animal fat are very tiny compared with the protein fat molecules but know how to live in the surface of the outer parts of the belly. These belly fats do not fight the large protein fat molecules and allow them to form a deposit on their outer surfaces. The reason why they don't fight them is because they feel protected when they land on top of them.

The protein fat molecules gradually accumulate in very large numbers and form a loosely attached colonies that make the fat in the gut of the belly. Those who eat a lot of protein nutrition grow large bellies and when their bellies extend outwardly they sag downwards because of the weight of the protein fat molecules. Belly fats are like foam and have numerous air pockets between colonies of protein fat molecules. That is why belly fats are normally very fine in the surface of the skin but blob when you press them slightly. Therefore the fat deposit derived from animals delight in the safety of the protein fat molecules and never want them to leave. The large protein fat molecules are not delighted to remain in the gut of the body and therefore explore more areas of the body to establish new colonies. They finally find a new appropriate location, on the chicks and around the face. There they delight because the skin pushes them closer and closer and form a more attached colonies of fat. That is why chin fat is more dense than belly fats.

When the person doesn't consume any protein rich food the level of protein fat molecules gradually dissipate. The outermost layer of the belly fat are no more desiring to remain in their location because no more protein fat molecules are available in the blood to form another layer and to make a new outer layer. As a result, the outermost layer of the belly fat depart and enter the blood stream to search for a better location. They search and search until they get to the face. They desire to cling on the outermost layer of the chin. The chin discards them because it is already saturated and cannot stretch the chin skin any more. The roaming protein fat molecules search the entire face and will not find any room for them anywhere on the face.

In their desire to make themselves attractive the roaming protein fat molecules restructure their shape. They become more round than the regular protein fat molecules. Still no room but now they have become like intruders who enter a house to rob with a mask. They leave the face and continue their roaming until they get to the outer surface of the small intestine. They begin their journey inside the blood stream as if they were newly formed molecules from protein nutrients ingested by the victim. From the small intestine they proceed to the liver. The blood arteries between the small intestine and the liver are where all the antibodies are concentrated. Not a single antibody exists anywhere in the entire blood system apart from the blood vessels between the small intestine and the liver.

When the disguised rounded protein fat molecules attempt to pass through the blood stream from the small intestine to the liver they are intercepted by the antibodies. The antibodies know the shape of protein fat molecules and understand that they are products from the stomach even if they don't have any value to the entire body. Hence, the antibodies don't devour them. Antibodies know all the sizes of the entire body cells, their functions, their secretions and their duties in the body. When they see the newly disguised protein fat molecules they immediately know that they are not desired in the blood system and identify them as foreign invaders. They immediately devour them. And so the process goes on and on. The outermost layers of the belly fat depart and enter the blood stream because they feel unprotected when they are exposed by the departure of the outer layer that covered them before.

As long as the victim doesn't consume protein rich nutrients for many days the dissipation of existing fat on the belly part of the body would continue steadily. That is why it is wrongly assumed that fat molecules are sources of energy to the cells because they dissipate very fast if a person doesn't eat any nutrients for a few days. Daily disappearance of fat deposit from the belly continues until the fat victim is reduced to extreme thickness. However, the face protein fat molecule deposits remain intact even if the victim empty of any protein fat molecules in the rest of the body. The face fats could only disappear in the victim's old age.

Then the fats disintegrate immediately the person reaches thirty-five years or older. The chin remains intact for a very long time but the other parts of the face become wrinkled when the fats disappear from underneath the skin. No one has ever had wrinkles on the face before the age of thirty-five years. If a victim remains fat after the age of thirty-five the chin fats don't disintegrate. The body fats will have to dissipate and disappear before the face fats begin the process of disintegration.

Everyone has his/her own unique way of absorbing fat molecules and disposing of them when they don't consume protein fat molecules. Obese people have a very high rate of accumulating fat in their bodies than regular people. They also have slower rate of disposition of the accumulated fat. Everyone is entirely different from the other. There is no one way of accumulating and disposing fat in the entire body. If you don't accumulate fat even though you consume a lot of protein then that is entirely unique to yourself and there is no one like you in the entire humanity and neither will there be anyone like you eternity. Hence, if you don't dispose fat as fast as a close partner who is eating exactly as you are, don't despair. It might take you a longer time that your partner but you will eventually succeed in getting the required amount of fat you want to get rid of. Even obese people could succeed in having a trim and thin body. It only takes them a much longer period time.

Many minds desire a thin body but delight in the accumulation of fat in the entire body. The reason is because fat accumulating mind thinks it is doing the body a favor. Accumulated fat is an ultimate protector of the entire body and daily accumulation of fat enhances that perception of the mind. That is why man is attracted by protein rich food because they are testier. Yam is tasteless and so is cassava. Wheat and corn contain a lot of carbohydrates and tasteless. Butter and cheese are the best source of animal fat and have no relationship to protein fat molecules. Butter and cheese are the source of energy to the functioning of the cells more than any other sources, even better than carbohydrates. Butter and cheese fuel the body of the cells because their molecules are loosely attached to each other and break down easily and release huge sparks in the process of atomic fusion taking place inside the cells. The energy they produce last much longer than that produced by carbohydrate molecules.

If you delight in eating cheese or butter you are always full of energy and eat very little of other foods during your lunch and dinner. Butter spread on a bread in the morning is equivalent to three times the energy you derive from bacon. A single butter slice will last for a whole day and the individual doesn't have to eat anything for a whole day. To derive a similar effect you have to eat three slices of cheese of the same size of slice. Hence, swallow a few slices of butter before you go to do construction work and you will never exhaust yourself for the whole day without putting any other food through your mouth and an entire shift. Marathon runners should only consume half a bar or preferably a whole bar of butter and they will run like a gazelle and will never show any sign of exhaustion during and after the completion of the marathon run. The Kenyans are great marathon runners because they can endure for much longer period of time than others. The main reason is because Kenyans, Ethiopians and Eritreans share the same staple food pertaining to butter.

Kenyans who endure much longer to others feed on solid milk with out extracting the fat content. So are Eritreans and Ethiopians. The reason why Kenyans, Ethiopians and Eritreans endure much longer is in the concentration of fat molecules in the milk they drink. Kenyan cows have much condensed molecules that Ethiopian cows. Ethiopian cows have a little more condensed fat than those found in Eritrea. And the cause is in what the cows feed. Grass in these areas are very rich in methane gas producing molecular structures. Milk is just a combination of five methane gas molecules combine with a single hydrogen molecule. That is all milk is and its extra content of butter.

Butter is mainly methane molecule combined with two or more hydrogen molecules. You can produce the best milk by just combing methane with hydrogen in a factory. Milk could be cheaper than water because methane and hydrogen molecules are not difficult to produce. Even people living in the Sahara desert could produce milk very cheaply. If you want more butter you combine more methane molecules to one hydrogen molecule and could have endless supply of butter. You can consume a ton of butter a day and it will have no adverse effect in the functions of any part of the body. Excess butter fats are deleted by the kidney with so much ease and their excessive numbers don't have any effect on its normal functions.

What happens to the mind that doesn't care about the protein fat molecules deposited on the inner surface of its blood vessels? Such a mind is deadly dangerous and the individual easily dies immediately the accumulated protein fat molecules reach the zenith of the tolerable heartbeat of the heart. Clogged arteries can only function normally by increasing the heartbeat dictated by the mind. When the maximum heart is achieved the mind clamors to reduce the heartbeat to a lower level. But by then it become too late. The heart capillaries burst their tubular passages and excess blood floods the surroundings of the heart cells. Almost immediately the heartbeat reduces to almost normal level. The capillaries don't repair and would continue to ooze blood and blood cells deposit outside the affected cells where the bursting of the capillaries had occurred. Ultimately many dark patch appears on the surface of the heart as a result of the death of trillions and trillions of hemoglobins. That is the first indication of a defective heart.

A defective heart does not kill a person because the surrounding broken capillaries of the heart cells don't die and their functions remain intact. The only thing is that their actions are drastically reduced. The heart remains at the normal rate of heartbeat and the person could survive for a very long time as long as the pluck deposit remains the same and no more protein fat molecules are produced by the individual. What happens to such a person if he continues to consume protein rich products. As explained above the process of the creation and accumulation of protein fat molecules remains the same. However, this time the deadly protein fat molecules disappear at a faster rate from inside the blood stream because they are designed to easily stick to one another easily. The newly created protein fat molecules don't start new colonies on the inner surface of the heart, or any parts of the entire body for that matter, when they could easily attach themselves on the already accumulated pluck on the inner surface of the arteries, veins and capillaries. That is why the heart beat increases, once again, because the blood flow is hampered more as the accumulation of pluck increases on that part of the inner surface of the heart.

The raising of the heartbeat is the only available directives the mind could issue because it has no other alternative. It can't repair the capillaries. It can't instruct the cells of the inner surface of the heart vessels to get rid of the then first deposits of protein fat molecules because of their enormous size compared to the heart cells. The mind is rendered helpless and desires no more protein fat molecules to be produced by the linings of the stomach. However, the stomach is overwhelmed by the amount of protein rich product the victim ingests through his mouth. The stomach goes into overdrive and produces so much protein reducing enzymes and reduces the protein molecules into glucose. Glucose has absolutely nothing to do with carbohydrate molecules and certainly not derived from carbohydrates as assumed by scholars of biology.

No one should be blamed by the erroneous knowledge of the process and ultimately the creation of glucose for they closely resemble a typical carbohydrate molecule. Knowing that. The cells are fooled into ingesting glucose to be used as energy producing fuel. However no spark is created in the cells' nuclear fusion for proteins are very complex compounds and do not easily disintegrate when oxygen molecule split into their atomic component which is the energy required to fumigate the food particles.

The finally accumulated protein fat molecules end up gradually totally blocking the affected blood vessels and no blood is pumped through that vessel. The mind doesn't know what to do. It desires to give the distressed and the starved cells deprived from food and oxygen as a result of the lack of blood flow entering into the dead capillaries. The mind doesn't want to increase the heartbeat to a dangerous level knowing that excessive pumping of blood through the blocked vessel would result in deadly consequences. The vessels could burst easily. As a result, the mind is rendered worthless and it know it. It just stays idly wishing it would get better. However, the heart doesn't give up upon itself and devises its own method of overcoming the major problem of a blocked major veins, arteries and capillaries.

The heart routes the blood flow of the blocked veins, arteries and capillaries to a nearby blood vessel just adjacent to the blocked one. The plan works and the heart saves itself from imminent destruction. However, the blood pressure of such an individual remains high and that is how high blood pressure resumes. As the routed blood reaches the affected cells of the heart it deletes its old arteries, veins and capillaries and creates new paths. However, the new path is not a blood vessel but just flood of blood around the dying cells that sent distressing ideas to the mind, that the mind ultimately ignored, and those that the heart took a unilateral plan to save.

The same applies to the status of a person who develops a little higher blood pressure. That wouldn't kill him and could live with such a heart rate for many, many years as long as he doesn't continue to consume even more higher rich protein nutrients. The rise in protein molecules produced the initial speedy heart rate. Additional consumption of more protein rich food produced higher blood pressure. Note that heart rates and heart pressure are two distinct messages. Higher heart rates are just the increase of the speed of the heart under a normal vessel of the vein, the artery and the capillary. However, high blood pressure is the measure of the heart pump to get blood through a rerouted vein, artery and capillary. It is very important that you distinguish it if you want to understand the deadly effects of protein fat molecules to the normal functioning of the heart.

What the heart desired is achieved and the person only experiences a suffocating pressure immediately above the center of his chest. He feels suffocated because the blood vessels in the heart have direct connection to the lungs. The lung is the first to be affected by the change in the high heart pressure and not the change in the high heart rate. Normally, such pressures in the center of the chest remain for a short time because the lungs adjust to the new low blood pressure that results from the blocked, arteries. Note that as the heart pressure increases the corresponding effect on the lungs is low blood pressure and less blood is pumped to the breathing organ as is the same in other parts of the body. However, the effect of high blood pressure is felt more by the lungs because they are basically blood vessels connected by a network of fibers, just next to the kidneys. If you eat a lot of fiber containing food your breathing ability is much higher than the person who doesn't consume any fiber rich nutrients.

When the mind realizes that the heart has been stabilized, not knowing how it took place, it remains silent if it receives distress ideas from the heart cells again. As a result, the mind becomes extremely insignificant and the heart is left entirely to itself to mend and repair any further defects in its functions to survive. The only thing is that the heart has no memory bank and implements an idea and doesn't have the capacity to remember how it did it the first time. As a result, when the victim continues to consume more and more protein rich products the stomach could only change the protein level that the mind asked it to convert to glucose. The additional protein the victim consumes are broken down to the large protein molecules and ooze into the blood stream through the porous interior surface of the small intestine.

The newly added protein molecules don't have to go far before they discover the final destination. They avoid the gut protein fat layers because they are so fragmented and very foamy and contain pockets of airspace between tissue colonies. Knowing that they advance to the face protein fat molecule deposit and they are not needed because the skin of the face could stretch so much only. So, the only alternative they have is the heart vessels where they could easily attach themselves to the protein fat molecules that had formed the pluck, which subsequently caused the high blood pressure in the victims heart and low blood flow in the rest of the body including the body of the heart.

When the newly formed protein fat molecules that were formed by the extra excess protein rich food consumed food, ultimately begin to rest on the already formed pluck in the artery it develops into a lump. This lump grows in size as more and more protein fat molecules are added to its out surface. Eventually, it totally blocks the artery and the heart again scrambles to save itself from immediate death.

It does something very crude that will cost it terribly. Instead of finding another re-router vessel of blood to feed the near starving cells of the maimed capillaries, it desires to expand the route that it created earlier when it saved the heart from imminent death. That is the cause of death of the victim. The expansion of the first route devised earlier destroys the main artery because the heart doesn't know how to increase its heart rate to exert the power needed to push blood flow into the second re-routed path to the affected cells.

The victimized person immediately realizes the re-routed path when his heart rate drops dramatically. He hold his center chest and feels as if his lungs are shrinking and chest cavity pressing inwardly. The mind doesn't know that there is a burst artery inside the heart. Hence, it responds to the distress idea sent by the lung cells. It quickly delegates its instructions to the ribs to increase the compression and expansion of its body parts. The person who experiences "heart seizure" ventilates at a very fast rate.

His lungs and not the heart causes the fast chest compression and expansion that results into a heavy breathing by such a victim of heart attack. The victim doesn't die from a high rate of the heart beat but of the opposite. The heart rate drops dramatically which reduces blood flow to the entire body first beginning from the low blood flow to the lungs. Hence, as the victim dies, he loses consciousness first, then his lungs collapse and immediately all the other body organs cease to function. When that happens, it is too late and nothing will bring back the person to life. However, if that same person receives aid quickly and the ribs are compressed and expanded at a low rate equivalent to the rate a person inhales and exhales in his lung, he could be saved.

When you do that, you are not starting the heart beat for the heart dies after every main organ dies. It continues at a very slow rate until it is starved of oxygen and then expires after all the organs are dead. The heart retains excessive oxygen throughout its existence because the mind knows that it is the most important organ of the body. It could survive without oxygen supply from the lungs for as much as sixty seconds at its normal rate and another twenty minutes in fading heart beat until it dies eventually. If a person receives aid and his chest is compressed and expanded at the breathing rate of an individual he could easily be revived because the heart successfully implements the second re-routing in a trillionth of a second. Compression and expansion of the chest cavity increase the lung functions and oxygen it transferred to all the entire parts of the body, including the heart. As a result, the saved victim will not have any defective part of his body and will function normally after the heart attack.

If the heart attack victim receives aid immediately nothing major will be altered in his life and will experience only fear of another heart attack. However, if that same victim did not receive aid with in ten minutes he is most likely to lose one body function or another. Knowing that what is the role of the lung in such a situation? You have to understand the mechanisms of how the lungs work to appreciate the above elaborate explanation of the deadly causes and repercussions of heart attack.

The ribs are responsible for your breathing. Without the ribs you will suffocate and die in ten minutes at most. The lungs are not the source of your breathing but the compression and expansion of the ribs. When the ribs expand there is a sucking effect in the internal space of the upper body. And the top part of the lungs are drawn outwards. When the ribs compress inwardly there is pressure on the top parts of the lungs and therefore the top part of the lungs nearly attach themselves to the bottom part. When the ribs expand the lungs become wider in their interiors and a vacuum is created. To fill the newly created vacuum, air gushes through the nostril until the space created by the widening of the interior part of the lungs is filled. That is how you inhale. The reverse occurs when the ribs compress and there is pressure on the top surface of the lungs. As a result, the lungs compress and the air that had previously filled the vacuum is forced to depart. And that is how you exhale when the air exits through the nostril.

Many people learn the art of CPR and are taught to assume that they compress the chest cavity to restart the heart beat. Nothing could be further from the truth. If it so, then why do you have to compress and expand the center of the chest cavity? That is not where the heart is located. Every dummy could have understood that but the writer knows that every reader of this writing is a fucked up moron. He doesn't delight in knowing that and would understand that if you would open up a little part of your dead brain.

If you want to know the importance of the ribs in your breathing mechanism consider a person with a cracked rib. Just a single rib. Why does he feel excessive pain every time he takes a small gulp of air. The reason is because he couldn't have taken that gulp of air without compressing and expanding the ribs. Every rib is relevant to inhale and exhale the slightest breath in your life. If one rib doesn't function properly then your breathing mechanism is severely disrupted. You will know that if you ever crack your rib. You can crack as many ribs as possible and continue to breath even though every breath is taken at the expense of great pain. However, if one rib is broken and snaps into two then you are doomed and the result is imminent death.

Now you know how the ribs are important and could associate the importance of the ribs in the creation of a woman as recorded in the bible. Eve was created from the rib taken out of Adam's chest cavity. The writer knows that some fucked up evangelists ignore this important fact and delete it as an error in the annals of the word of God. However, the bible is the word of God and none of its content is irrelevant. Murders evangelists couldn't figure out the relevance of this simple functions of the single rib. Now that they know that man could not survive a day with a single broken rib how much more if it is out totally which is an imminent death. A woman doesn't have that problem. A woman could live if any one of her ribs are taken out permanently.

Not a man. Figure it out in your demented and porous brain cells. Man can live without a woman but the reverse is true with women. A woman can't live without a man. That is the truth in the past and will remain the truth forever and to eternity. Having known that what the hell is going on with men and women? We know now that a woman is just a rib and has to find a body to feel like a complete woman and that is why her mind never ceases to think of her husband. That is all.

A woman could love a man with her heart and soul just like she would love her own loving mother. But no man has ever loved a woman like his dear mother. It never happened and will never happen to eternity. A woman could live with an abusive and fucked up murdered of the mind for a hundred years if they live that long. The woman can't see herself without her soul mate which she sees her husband to be. Every woman desires to have a complete body and that fact could only be achieved if she marries to a man. Lesbians are just two fucked up deadwoods and degenerate subhumans.

When a wife begins bickering against her husband there are two things happening. One, she is desiring to dominate him and second she is desiring to discard him. Both of the desires are detrimental to the liberty, freedom and pursuit of happiness rights of the man. The man desires to stay and live in peace but the woman continues bickering and turns his head upside down. The man desires to search for his family help to rectify their differences. That is what happens in developed cultures and not among white people because they are fucked up, demented and subhuman men.

That is not all the disgusted husband does. He seeks psychological therapy and he is the first to respond to such a treatment. The witch wife doesn't respond equally and only suspends her weird desires for a time. She never forgets her real intentions to either dominate or to discard the husband. The demented sub human husband truly believes that their marital situation has been repaired and fools himself and assumes a lifelong happy life. That never happens. A determined woman is more evil than the devil and even Satan is amazed by her level of cruelty.

The next stage of the woman's sickly mind and memory is to kill the man permanently. The woman knows exactly how to do it. A foolish woman will blow the head of now dying carcass with a single bullet and claims every fucked up reasons to justify her actions. In some fucked up cultures that is considered a justifiable homicide and the fucked up witch and murderer rendered scot free and even considered as a hero. In such a society women are the most abused and men feel heavenly power on them and think they are the replica of God instead as the image of God that they are.

Only white people think they are the replica of God in the whole world. Every other culture in the world accept their subservient to God. That is why there are no Chinese supremacists, African supremacists, Native Indians supremacists and so on and so forth. If you want to see the most respected woman on earth go and see a Mensa woman and how much she is respected and revered by her husband and children. Similar cultures abound in all the parts of the earth except the white people.

White people are the most inferior people when it comes to the evolution of cultures. White people gathered together and lived as a group people not in a long distance of time and their culture is still evolving. The Chinese husband and wife live in perfect harmony more than any other societies of the world because their culture evolution began far more than the other cultures in the world. That is it and there is no other magic. Women in China have perfect knowledge of their freedom, liberty and know how to pursue ultimate happiness by far more than white husbands and wife. That doesn't mean that there are no wives that have the devil's design in their minds and memory but they are very few compared to European descendants wives who are by far the most wicked, evil and mind and soul murderers and their sickness destroy their husbands permanently. Then what is happening inside the fucked up mind and brain of sub human white husbands?

White husbands think they are superior to other races but don't know that they are inferior to their wives. They think they are superior to other races not based on silly ideas but facts. White people are indeed superior to other races of the world because they once ruled the world and dominated every society and still do until today. Hence, a white man is not fabricating his God given freedom and liberty and his desire for ultimate pursuit of happiness. The white man did not deprive anybody in the countries they dominated and still dominated of their ultimate freedom and liberty and their wishes to attain maximum happiness.

If the people they dominated and dominate now are fucked up fools to see them as superior then it is not the fault of the white people but of the sub humans the dominated the people assume that they are. White supremacy is evil when it infringes on the freedoms and liberties of those upon which it is enforced. Otherwise, there is absolutely nothing wrong with white to feel superior to other races. Their memory storage bank is filled with so many ideas that are facts and white people will never relinquish that position in the hierarchy of their priorities in their minds. Their children will read the history of the white race and fill their mind with such factual ideas and as a result white people will perceive themselves as superiors to eternity unless something very drastic takes place to dispel their right concept.

Going back to the white husband and wife let us assume their names are Jenny and Martin. Let's assume Jenny is a fucked up and deadly mind murderer that represents a typical white wives that they are. Lets assume Martin is the sane and white supremacist with no silly ideas that represents most of the white men that they are. Jenny doesn't stop plotting to destroy Martin from the day she deceived him to marry her and the foolish Martin thought he was marrying a perfect wife and dreams of factual ideas in his mind's memory that a typical normal human should perceive for that is what white men are. But Jenny the sickly witch instantly, after the day of the marriage, assumes Martin to be her personal guard that a typical woman assumes her husband to be in the entire world.

Jenny eventually forgets that she is a vulnerable woman and begins to assume to personal protection when she realizes that Martin will never give his life to protect her. No white man will die for his wife and that is a fact and that is not the normal God given instinct of the role of a typical man all over the world. Go to any other culture in the world, except the white race, and provoke the wife see the reactions of a husband. In some extreme societies you could end up dead for simply insulting the wife of an honorable man.

The moron Martin doesn't know that his wife is sickly and deadly murderer of the mind that most white wives are. But the white woman did not become wicked and evil overnight. It is a process that took place from the time whites began to live collectively, which very recent compared with other more developed cultures. All men and women in other cultures began the same way and passed through the stage of evolution the white race is today. All cultures treated women as animals for many centuries. Over the millenniums those cultures evolved gradually and formulated rules and laws governing wives and husbands to coexist in perfect harmony. The white culture is not developed and not even to that of Chinese culture or Japanese and not even many African cultures. In essence, the white race is the most inferior race and is at early stage of human mind development.

What happened to Jenny the witch wife of the saint Martin? Jenny never loved Martin from the beginning and will never love him ever. No white wife loves her husband like her mother, which is the ultimate a human could render. It never happened and will never happen for another millennium unless something drastic takes place to change such a solid perception entertained by the minds of white wives. Jenny fucks Martin like any other woman and derives the same sexual pleasure as any other woman of any culture in the world. The only problem is that Jenny is a free-minded woman and never knew that she was sexually dead before she even met Martin. Martin never knew that his wife Jenny is just a log of wood in bed and doesn't know that she fakes her orgasm, as most white women are. White women derive pleasure from sexual intercourse with a man and are normally loud in expressing their fake pleasure and don't know that they are sexually dead monkeys.

When any woman grows to maturity at the age of thirteen she is ready to mate and should get married at that age which is most preferable. Many cultures realize that fact and ensure that their daughters are married shortly after that rife age. White culture does not recognize that fact as the truth and hence, their daughters marry at a much older age. Nothing wrong with that as long as the daughter could keep herself as a virgin. You will understand the importance of virginity when you finish reading this topic and don't start filling your fucked up and demented white brain and mind with silly ideas. Just keep on reading if you have at least a dead mouse's brain.

When a virgin has her first encounter with a male partner the cavities in her virginal react simultaneously when the penis is inserted at the outer surface of the virginal tube. As the penis penetrates further and further into her virginal tube the cells in inner surface of the virginal tube register every characteristics of the structure of the penis. Not only that the texture of sensitive head of the penis are only unique to that individual and no human has ever had such a texture since the creation of man and there will no one like it to eternity. Look at your face on the mirror.

Know that no other human has ever been created that remotely resembles you and there will be no one that will have like your face for ever. Just compare that to the sensitive head of your penis and that how unique the characteristics of your penis are. However, twins could have similar face and other features as well. However, with twins when you consider their entire body structure you realize that they are different in a big way. Not only that twins don't think alike even if they imagine very similar fantasies. That is why twins could easily discern what the other is imaging fantasies. Otherwise, they don't think, imagine and remember through their minds as one being.

So when a girl encounters her first sexual experience, the inner cavities of virginal tube recognize and could draw an exact replica of the inserted penis if they were artists. The girl experiences a little discomfort and pain at first but after a few strokes delights in the sensation created by the friction of the penis head on her relatively rougher inner surface of her virginal tube. The next day, the same thing happens. The only difference is that the moment the head of the same man's penis touches the outer surface of the virginal cavity the sensory organs instantly recognize not only the texture of the head but also the unique odor that is emitted by the entire penis of the man. That again is unique only to that man and no one else on earth.

So the woman mates with her new man and eventually she becomes pregnant and they have children and live a happy life thereafter. Every time the man inserts his penis in her virginal tube the now familiar image of the man's penis head is readily accepted by the sensory organs of the outer surface of the woman's virginal cavity and she gets maximum pleasure and squirts virginal secretion every time she has an orgasm. Orgasm is reached by the woman when the sensitive virginal tube surface causes a small friction when it rubs against the man's penis head. That is why penis size has no significance in sexual intercourse.

A Chinese woman is just as good to reach orgasm as a woman in Armenia or elsewhere. The virginal tubes of any woman of any race or color is a tightly compacted elastic membranes. The virginal canal is so narrow that it could only allow urine to flow under very heavy pressure exerted by the bladder. Its passage is not even as wide as a simple ball pen. Those who have larger penis size have no extra advantage over those with smaller penis. A woman who had sex with a black man the first day will have the same sensation if she had sex with a Chinese the next day. The only difference is the amount of stretching the virginal tube had to do and that has nothing to do with her process of reaching orgasm and is of no significance in the pleasure she receives.

So, lets assume Martin is the one who married the virgin we discussed above. Jenny's virginal sensory organs recognize martin's penis and everything is good and dandy. Every time Jenny screws Martin she squirts and reaches orgasm in a minimum of two minutes and that is very healthy and nearly appropriate. After many tries Martin endures for one more minute and three minutes is all Jenny needs to sustain her pleasure. Knowing that what happens if Jenny tries to experiment and flirts with her co-worker and decides to mate with him?

When the co-worker whom we shall identify as Mike enters into a confined space with the whore Jenny she immediately develops extra sensations more than that of her experience with Martin. Mike tells Jenny that is not appropriate to mate for she is married and is also one with two children. But Jenny craving is uncontrollable and will not leave without mating with the adulterer Mike. So they decide to mate.

Mike begins to insert the head of his penis into her virginal canal. However, the outer cover of the virginal canal does not recognize the texture and smell of Mike's penis. It therefore alerts the mind and an idea is created instantaneously. The mind sends the idea to the brain memory and the brain memory scans its entire storage using its search engines and comes up with an exact picture of Mike's penis head. The message is sent back to the mind and it searches its memory storage bank also and finds no information of Mike's penis. Hence, the mind alerts the entire reproductive organs and dictates different instructions to them.

The outer surface of the virginal canal produce pungent secretions to ward off the intruder, Mike's penis. The virginal canal tightens to prevent entry of the foreign object. The virginal secretion, situated between the cervix, shut off preventing any squirting that produce climax. The cervix that house the eggs produced by a fertile Jenny seal their exit doors to prevent her eggs from flowing into the virginal canal. All these take place in a millionth of a second.

So Mike eventually inserts his penis and does his stroking and climax but Jenny could not squirt and would not attain orgasm. Mike is a sexier than Martin and lasted more than four minutes but Jenny couldn't climax. Eventually, Mike would tell her how much he enjoyed sex with her but Jenny would be disgusted with Mike and would swear that she will never mate with him again. Jenny was not satisfied sexually even though Mike aroused her fully for he spent a lot of time caressing her nipples and rubbing the surface of her virgina and other vital body parts. So, immediately she enters her house after work she desires to mate with her saintly heart Martin. Martin does the normal things he does and eventually decides to penetrate her. This time, however, the outer surface of Jenny's virginal canal has a steady flow of pungent secretion intended to ward off that identified Mike's penis as a dangerous foreign object.

When Mike mated with Jenny and finished with her the perception of the inner surfaces of the virginal canal is utter confusion. The sensation nerves on the inner tube of Jenny's virgina could only respond to one penis head size in their entire history. They don't have their own memory and therefore see Martin's penis head as another foreign object. And the mind reacts in the same manner as when Jenny had sex with Mike and alerted and shut the entire reproductive system. Martin reached climax when he ejaculates but Jenny would not reach the same orgasm she was used to when she mated with Martin before. And she will never reach orgasm for the rest of her life and neither will she see Mike again.

Wait a minute, that is not true. A woman that had an affair once doesn't stop producing children. How could that be. If so, no white witch will have children for they nearly all of them have had more than one sexual partners. How screwed up could the writer be will assume the dead heads of the white race. The writer knows the mind of every man and knows when a silly idea is created and how damaging it is to the wellbeing of an already fucked up brain of white moron. So the writer explains further more and learn to decipher the truth of your already hateful against the writer could endure. If you stop reading any further than these words, fuck you. You are already screwed up anyway and some of the white fucks are beyond redemption, not even the mind-of-man could heal and save them. Many whites already believe in their subhuman status and are desperately attempting to establish equality with the other and more developed cultures of the world.

The above state of Jenny's body is true if Jenny mated with her husband, Martin, the same day. She becomes infertile immediately and remains so for life. Lets assume the fucked virgina of Jenny did not see the essence of further mating with the still saintly Martin on the same day. Jenny formulates whatever excuses to avoid having intercourse with Martin for her cunt is very sore because Martin is a very sexy man. So what happens if she had sex with Martin the next day or thereafter? If she had sex with Martin the same day she committed adultery her ovaries produced a thick film preventing any fertile eggs from exiting the ovaries ever and that is why she would up as a barren woman.

But if she didn't have sex with Martin on the same day but did so after one, two, three etc. days the ovaries would not seal their opening gate to the center of the virgina cavity permanently. Even though the ovaries of Jenny produced a simple seal over the gate to prevent her egg from migrating to the open field of the virginal cavity, such a state resulted for a short period of time. So, when Jenny mated with Martin the next day her ovaries were not closed for the sealing lasted for only a few hours. That is the difference between mating immediately with soul mate after an adulterous adventure and doing so after one day. Big and deadly difference!

Why should a day difference of mating with adultery produce such a widely varied difference? You remember what happened to Jenny's reproductive system after she had an affair with the adulterous Mike. Two prominent things happened. One, her ovaries shut down with a thin film of its exit canal. This thin film disappeared after a few hours when Jenny's sadist mind discerned that there was no danger to the productive system when Mike pulled out his penis from Jenny's sick virgina. Hence, the fertility of dirty Jenny was restored. Number two, Jenny never squirted when she mated with degenerate Mike even if he tried madly to please her.

She never reached full orgasm at that time and she will never, never, never ever see another orgasm to eternity. So adultery destroys any woman permanently and no adulterous woman will ever enjoy having sex with a man from the day she deviated from the mind-of-man. That is why adulterous women are stoned to death in some developed societies. The fucked subhuman European and his descendant doesn't know that his witch wife is a deadwood. He might as well fondle the head of his penis instead of fucking fat Jenny for that is the end result of a fucked up bitch.

What happened to the witch Jenny when she mated with saint Martin the second day or the third day or the fourth day etc. After she mated with deadly Mike. The same thing happened as she experienced when she was initially disvirgined. The outer surface of her virginal canal composed a perfect picture of saint Martins penis and the outer circumference of Jenny's virgina identified saint Martin's odor. Every time saint Martin attempted to insert his penis into the constricted circumference of Jenny's virgina it smiled and expanded widely to facilitate easy access to Jenny's virginal canal. Saint Martin inserted his penis with ease and stroked repeatedly until he climaxed. That would be true if Jenny mated with saint Martin only for the head of Martins penis produces a secretion to lubricate the inner surface of Jenny's inner virginal surface. That is why the head of the penis shines all the time.

The head of the penis secretion is not smelly, however, the body of the body, from the base to the shinning head of the penis, has a gas that is unique to that penis and cannot be replicated by any other body of a penis throughout the history of mankind and never will be. When this gas combines with lubricant of the inner surface of the virginal canal it becomes very pugnacious. Again this newly formed deadly combination is unique to the gas emitted by saint Martin's penis body and the sickly dead witch Jenny. There will be no such combination in the entire humanity and forever. Only Jenny and Martin could produce such a unique combination and will have its own unique smell and that is why the deadly sensitive outer circumference of Jenny' clitoris recognizes Martin's penis.

Hence, if a woman knows only one man in her life she is guaranteed to have maximum pleasure from her husband's sexual intercourse and will produce up to twenty child, if she so desires, and will not be relieved of her freedom and liberty and the pursuit of ultimate happiness. Adultery is very deadly to a woman's pride and happiness and destroys her permanently. Women who delight in multiple sexual partners end up deader and deader with every fuck they have with sickly consequences. A one-time adultery only kills a woman's ability to reach orgasm but could stay fertile until a maximum of thirty years. After that she becomes barren and cannot have children. A woman who mates with more than five men in her life her fertility period is only twenty-five and becomes barren after that period of her life. This is the ultimate truth giving one or two years up and down. A prostitute that mates with more than three men in a day cannot produce children forever and ever. Now you know what the fuck I was talking about. She is just a mule.

What happens if a man becomes an adulterer? You think absolutely nothing but has its own consequences too, even though not as severe as a woman who has extramarital affairs. But that doesn't mean he remains a stallion forever. The first time a man mates with another woman other than his soul mate he discharges his semen on the virginal canal cavity of the other woman. Nothing drastic happens to the penis of the man and doesn't suffering anything in his manhood. Nothing would happen that way and goes back to his wife and never mates with another witch again.

That same man is tempted by his sexual adventure and sexes another witch different from the first adulterer. Then something happens. The second time the man mates with a second woman other than his dear wife the head of his penis discharges a film of secretion varied with the odor of the secretion of the second woman's inner surface of her virgina. That is not bad, only a different odor results and does not harm any part of the penis. When the man mates with the second hole apart from his fine wife, the same process takes place.

Remember the adulterer husband produced a different odor when he mated with the first witch. That odor becomes the new identifying smell of that penis and the secretion of the head penis changes to produce the new smell.

When the dirty scum mates with his soul mate whom he married when she was a virgin the wife's outer clitoris does not identify it as that of the husband and constricts tightly barring entry into the delicate virginal canal. Eventually, the husband penetrates his wife and mates with her. But before that all the reproductive system of the saint woman shut down just as Jenny's when she first mated with Mike. As a result, the squirting nozzles of the sensation producing canal to reach climax shut down forever and ever and the all time virgin woman will never experience orgasm throughout her life and will remain fertile only for a maximum of thirty years of age. The last result is divorce because the deadly man imagines that his wife is cheating on him even if he is the one who inflicted deadly blow upon her. He murdered her.

If the man continues his sexual endeavors and mates with many other women as fucked up niggers do. They have relatively bigger penis than, for example, Chinese, however, their perfect erection lifespan of their penis is by far less that the Asian. For example, every time the brain dead nigger mates with a new woman, the odor of the head of his penis changes and has difficulty penetrating the woman even if the woman is a declared whore. By now you know what happens to the entire reproductive system of a woman when her virgina encounters a new unique odor of a man's penis. No need repeating the process for it is the same with all women and the resulting outcome the same also.

The daft nigger sees himself as a stallion because that is his DNA and acquired it as a descendant of a slave mentality. The white supremacist slave master chose a stud from amongst the slaves he owned and gave the damned nigger the only right to mate with the damned nigger women. The slave master's intention is to produce stronger and more productive children of slaves to serve him and increase his wealth. The fucked damned slave thinks it is his birthright to mate with all the women in the farm. And that is how the diddle nigger thinks today. His grandfather was a fucked up death imparting and murderer of women, and so was his father, and so he is today. Fucked up brain.

So what happens to the penis head brain nigger who delights in implementing death sentence on innocent women, blacks and whites? His mind never determines to kill women but that is exactly what he does. How many does he have to kill before his murdering spree is terminated? Knowing that the dick head nigger desires to fuck white women on their rectums knowing that white women delight in doing that. So he searches and searches until he finds one. The dick head nigger has no shame and would do anything to find his prey. Normal people would despise themselves if they are repeatedly rejected by two or three women within a short period of time. But the fucked up nigger would not relent his search to kill white women even if he is rejected fifteen times. He will nag and nag until the white woman succumbs to his desire. Ultimately he will find a willing Jenny who desires to experiment with a black man. So he fucks her up on her ass and permanently tears and disintegrates her tightly constricted rectum. Jenny's rectum never recovers its elasticity and never again becomes airtight that normal peoples rectums are.

When you excrete waste matter from your large intestine through your rectum a lot of pressure is required to exit the airtight constricted membrane of the outer surface of your rectum. When one has serious constipation and desires to empty his bowels his mind tells the rectum to expand its muscles to accommodate the larger circumference of the waste matter. As a result, when the waste matter nears to exit the inner part of the rectum, part of the large intestine follows the soil. As a result, none of the larger circumference of the shit touches the outer surface of the rectum. The reason is that the outer part of the rectum is constricted muscle and when it expands it is very rough which causes a lot of friction when in contact with the exiting shit. That is why a small part of the lower large intestine closest to the inner part of the rectum exits with the shit. That way, the shit slides out with little resistance from the rough outer surface of the rectum which could have drastically reduced its discharge rate.

So Jenny doesn't have any outer muscle of the rectum. The diddle head nigger had permanently obliterated it with his bombastic dick. What would happen to Jenny as a result of that? Whenever she tries to go the toilet she only has loose stool. She wonders why she was always having diarrhea. Even more she would begin to see small markings on her pant not knowing where they came from. Finally, she could not hold her water stool for a long period of time. Going to toilet becomes a torment. She runs for the toilet every time there is a slight pressure emanating from her large intestine. She shits two to three times a day and all the time watery substance. So she fucked her ass to smithereens.

Hat happens to the nigger who inflicted the deadly blow to Jenny's ass? His large penis desires to fuck anything that has a membrane and has a hole. The degenerate nigger never realizes that he has dealt many such death sentences on so many women. So he desires to continue doing so. That is all he thinks and that is all he knows. But something happens that will hamper that deadly desire. The head of his penis begins to produce small boils around the circumference very close to the body of his penis. He wonders why he has such a disease. But that is not a disease but a reaction to bacteria unique to the large intestine. These bacteria are not deadly and don't destroy any part of the body. However, their secretions are poisonous and produce odors so pungent that will blow the sensory organs of the nostrils. So he wants to know who gave it to him. He fucked other women after Jenny and he begins to hunt the last one he screwed. He beats the hell out of her and is put in jail for attempted murder.

In prison he desires to masturbate for he desires to mate with women all the time. But he can't reach a full erection and doesn't climax even one time. So he decides to fuck inmates in prison but he can't find any willing one. So he is frustrated and wouldn't know what to do. After many tries he finally finds an accomplice in prison. He attempted to fuck him but his dick wouldn't stand. He tries and tries without any success. So he gives up and his cellmate desires to have sex with him as a woman.

The dick head nigger had never had his ass penetrated before so he rejects the request of his cellmate. His cellmate is agitated and he fucks him up with words. The dick head nigger doesn't like to be insulted so he grabs his cellmate and strikes him so hard that he bleeds profusely. The guards respond and place the dickhead nigger in the hole and spends a lengthy period of time there. When he is released he does not relent and attempts to find a partner to sodomize him. He find one and tries the second time and he doesn't reach a perfect erection.

The cellmate desires to fuck him also and this time the dickhead nigger relents. So his cellmate penetrates him and destroys the outer membrane of his rectal cavity. So the dickhead nigger reaps exactly what he sowed and the death sentence he passed on Jenny is on his head also. So that is what happened to the dickhead nigger and never would he shit solid matter again all the days of his remaining life. So, what is the lesson derived from the above story?

Jenny will never restore her rectal muscle again forever and ever so also the dickhead nigger. Not that they will never pass excess gas through their rectal cavities. Gas in produced in large quantity in the large intestine. The season why gas in produced in the large intestine is to create a barrier between the feces and the inner parts of the colon. Gas is preferred by the mind because lubricating the large intestine with enzymes or fat deposits would hamper or block the flow of vapor into the blood stream placed immediately below the thin film covering the inner part of the large intestine.

When gasses are too pressured they desire to exit. Hence, when the pressure built on the side of the stool is required to exit it pressures the rectal part of the body. The rectum is airtight and does not allow any gas from exiting and certainly not stool. Therefore the mind tells the rectum to open a hole equal to the opening of a syringe. And the person releases a squeak powerful enough to crack the shell of an egg. If the gas pressure was too excessive the mind directs the rectum to open an exit equivalent to two openings of a syringe. This time the person makes a sound of a blob. That is how the rectum works and very relevant to the final process of producing a perfect and circular rounded stool.

Jenny cannot blow a fart and neither can she ever hiss gas through her rectum. Neither can the dickhead nigger. Every time they desire to desire to go to the toilet their watery stool is so explosive that soil particles scatters all over the toilet. The reason is because the gas ordered by the mind to protect any friction between the stool and the inner surface of the large intestine cannot implement any of its intended functions. Hence, instead of accumulating on the side of the stool it pressures the waste matter from above of the watery stool. Jenny can't hold her excreta for long because the pressure on the watery soil of her waste matter is too much. That is how Jenny and the dickhead nigger will live for the rest of their lives. And that is not all.

When Jenny shits in a public toilet she blows her blasted soil all over the bowl. She cleans um herself but uses much toilet paper to weep herself clean. In the process, she drops a tiny speck of her excreta on the surface of the toilet top. Another person uses the same toilet and sits on the tiny, almost invisible, speck of Jenny's stool. The person doesn't realize that he/she has been contaminated. A day later the infected person develops watery stool and will be a replica of Jenny and will never see a sold stool forever. Many people infected by people like Jenny never recover for life.

What about the fucked up, women murderer, and dickhead nigger? Not the same thing as that of Jenny. Jenny is a woman and produces the deadly bacteria that is not present in a man's rectum. Only women. The dickhead nigger doesn't know what to do when he blasts the toilet with his soil and does everything possible to wipe out the spread particles all over the toilet.

Eventually, he succeeds. Unfortunately, one little speck of his excreta clings to the surface of his hand. That is how he infects other inmates when he just passes near them. He doesn't have to shake hands with any of them. The bacteria in a homosexual is airborne. It is very infectious and deadly for reasons you shall see. Jenny's feces produces water stool on those she infects but the fucked up dickhead nigger murders a man that comes close to him. When you finish reading this piece of information you should run for your life whenever you see a faggot as if a sharp knife-wielding madman is chasing you. Don't you ever shake hand with a fag.

So what happened to the dickhead nigger inmate? What happened to the many he infected while incarcerated in prison? As Jenny, the dickhead nigger blows out his excreta all over the toilet also. Those infected because the pores of the thigh are very wide and the bacteria gain easy access to the blood vessels and thus the watery stool imparted on the unsuspecting individual. But that is a child's play compared to the dangers poised by the fag dickhead nigger.

When he mated with his cellmate, the dickhead nigger did not delight in the penetration of his anus. However, his cellmate ejaculated and cleaned up his penis afterwards. The cellmate doesn't become infected. But the dickhead nigger is done for life. The sperm in his anal cavity is reduced to a deadly protein molecule. That doesn't happen in Jenny's anal cavity. The protein molecule formed from a combination of his cellmate's sperm and the secretion of the inner surface of the large intestine is what makes the newly formed protein molecule very deadly.

The protein molecule of the sperm combination travels through the blood stream to the brain. It doesn't pass through the digestive system, as the normal proteins are formed, and such proteins not intercepted by the antibodies between the small intestine and the liver because they are food molecules. However, the newly formed protein is a deadly one and if it passes through the antibodies it would be neutralized instantly. Hence, it clings to the brain membranes on its first trip in the blood nearly all the time.

Those newly formed proteins, for they are in millions, that flow into other parts of the body are forced to go through the said antibodies and are permanently disabled and urinated. However, a few find their way to the memory brain and stick to the cells. The memory brain sends a distress wailing idea to the mind. The mind sends the same message to the memory brain to be scanned and assessed. The ultimate resolution is returned to the mind. The mind scours its memory storage bank and concludes that there is a very grave danger inside the memory brain which could cripple the entire sensory system.

The mind senses that the newly protein molecules were formed in the large intestine and analyzes the content of the molecule and finds a sperm factor. So it delights and knows the dickhead nigger is a female because only women find sperms in their virginal cavity. However, the mind is confused because it represents a man not woman. So it scans its memory storage bank again to confirm its initial findings. This time, it confirms that the man the mind thought was representing was indeed a woman.

So the mind issues all alert to all the entire body parts to change everything that represented manhood and correct the error to that of a woman. The mind discerns that the dickhead nigger does not have a virgina so it attempts to open a hole directly under his testicles to change him to a woman.. The mind also finds that the dickhead nigger does not have cervix and only prostate. So the mind kills the prostate immediately and designates it as a dangerous invader. Nothing happens to the prostate except that it does not absorb the toxins in the blood that it is designed to breakdown and discharge into the blood stream as safe and nutritious food to the cells of the entire body. Such toxins could only be reduced to a safe level by the prostate and no other organ of the body including the kidneys, the liver and pancreas could perform such a vital function for the wellbeing of the entire body.

The next stage is to form nipples that are absent in the dickhead nigger's body. Next is the curvature of the body. Next is the secretion of body odors for men have different odors as women. Next is the eyelids and eyebrows. Next is the dentures. So it goes on and on until all the entire body parts of the sissy fucked up mama is transformed into that of a woman. The dickhead nigger that has been transformed into a fucked up mama would plead with inmates to fuck him all the time. The inmates refuse him and is frustrated. All the time the fucked up mama sits by herself, for she now knows she is a woman, and reminisces all the things she did to numerous innocent women like her. She desires to tell them that she is sorry and weeps some time in her cell. The cellmate rejects her and will want to change partner and he succeeds because the fucked up mama weeps all the time and is distressed. That is how the dickhead nigger becomes a sissy mama.

The new protein cells in the membrane of the memory brain cling to it for as long as they can and then detach without causing any harm to the memory brain cells. They enter the blood stream and are eventually devoured by the antibodies and urinated when they pass between the small intestines and the liver. The ugly sissy mama would be released from prison after serving her term and looks for men to fuck her. She delights when she sees women but her dick is disabled permanently and reduced in size to that of his infancy. He can't have erection forever and ever. That is why men want to have sex change operations to calm their anxieties and to fulfill their womanhood.

As for Jenny she endured for some time squirting shit on her underwear for a few years. Eventually she didn't know what to do with her open rectum. She desired to mate through her ass but men became disgusted because their penises came out totally soiled after sodomizing her. They insulted her and some times beat her up. Eventually, her anus would become porous and shit would ooze in large quantity into her underwear. Her only option is to wear women diapers to contain the growing flow of soil into her underwear. Soon, that wouldn't be enough to guard and preserve the seeping shit into her thighs. That is how the life of Jenny would end up. Eventually, she would admit herself into a nursery home at the age of thirty-nine and would live there until she dies. Women of seventy or eighty experience such a discharge and men of ninety. But Jenny destroyed herself with just a simple curiosity and a single incident of adultery.

Now that we know what possible scenarios could take place in the life of an adulterous wife, what happens to an adulterous husband? When the mangled mind white husband married a saintly white wife he assumed a perfect life thereafter. To clarify their hateful lifestyle lets assume the wife's name is Amanda and the man's name John. John desires deviant ideas in his brain. For obvious reasons explained above he sees himself as a superior human being even though he desires to remain equal to all family of man. He doesn't know that he has a white supremacist tendency and never neared blacks especially because he knows that blacks are murderers, thieves and dickheads. He has a point there and it is a fact.

John desires to love his wife with all his heart and soul, and he does, and that is why he married Amanda in the first place. John is a Pentecostalist and prays all the time. Amanda, his saintly wife, also prays with him, and even more, for she is always in touch with her inner mind with the Lord. They were both virgins when they married for their church advocated abstinence before marriage. Amanda never knew that her madly love for John will ever fade away and so also John's love for his wife. Both became madly in love with each other immediately after their married and after the first sexual interaction. So they thought all the people that attended their wedding, including their parents, who thought they bred a healthy, vibrant and loving couple. So everything was perfect and the world was at their feet. Nothing major happened in the first month of their relationship and there should be nothing more for the first month is a sacred month and a period of great temptations. That is when couples know whether their marriage will last or not.

Something takes place after a few months that disturbs Amanda's mind. John wants to lick her cunt and that repulses her but she finally relents because she did not want to discard John's sexual fantasy. She gets no pleasure except John's saliva on the outer part of her virgina. So things continue as normal and the married was destined to last for a long time. Soon, John asks for a different fantasy.

He wants Amanda to suck his dick and Amanda resists vehemently but John shows her pornography video for that is where he got his fantasy from in the first place. Amanda sees repeated sexual acts and the women performed blowjob on all the men all the time. So she imagines that it is the right thing to do for she assumes that all women sucked the penises of their husbands. So she performs her first blowjob on the sofa and John responds more eagerly to make love to her. And that is okay and dandy and no harm on either of them. If that is what keeps her husband happy, so be it.

However, Amanda watched the pornography video repeatedly and wondered why some of the women had anal sex. She resented the idea but saw almost all the women express ecstasy when they were penetrated through their anus. It wouldn't be a long time before John makes such a demand. Even if Amanda resented the concept she didn't see the harm and she suggests it eventually and John rejects it vehemently. Amanda becomes a vixen and assumes she is a jilted wife and that is how John's mangled mind begins to murder his marriage. John assumes anal sex is a sin while Amanda, his rib, is a sinful woman. Very soon they begin to have an argument intermittently. John doesn't know what is going on but Amanda is specific with her desire to experience anal sex, which is a viable demand, for John desired blowjob and she consented in order to please him. Amanda feels John has betrayed her love for the bible says that husbands should please their wives to the maximum.

So, normal life continues and both earn enough money to pay their mortgages with much extra money to handle other expenses to support their joyful life. Not many days later, Amanda gets pregnant and delivers a healthy baby and names him Jesse. Immediately, she gives birth to a lovely daughter and names her Melissa. And two years later they produce another healthy girl and name her Megan. Yet gain, two years later another baby is born and they name her in memory of Amanda's grandmother, Heather. So Amanda begins to take birth control pills to avoid further pregnancy. They continue to have healthy sexual experience for a period of time after the fourth child was born. Amanda doesn't know that her husband is distancing himself away from her and desires to cheat on her.

So John finds a beautiful blond bimbo on his way home from his work and desires to fuck her. He is disappointed when he finds out that she is a prostitute when she demands money in exchange for sexual favors. He delights, however, because his escapade will be kept secret for the prostitute will not want to know if he is married or has a child. So he bargains with her. The prostitute tells him the rate for each sexual fantasy she could perform on him and finds out that bottom fuck is the cheapest he could dish out without disrupting his scanty budget.

So he fucks the prostitute on her buttocks in parking lot and then heads home feeling as a satisfier of his fucked mangled mind. So, he makes love to his wife the moment he enters the house while the children are watching TV. He does that to waylay any suspicions from Amanda for he was an hour late from his normal routine. Amanda never noticed his coming late and did not inquire why her husband something unusual he never did before. Amanda loves him more and assumes everything is blissful and dandy.

John sees the same prostitute a month later and picks her again. He got away the first time, and what the heck, he could do it over and over gain and his wife will never ever find out. So he delights in anal sex but never tells his wife for revealing such secret will prick Amanda's inquisitive vixen mind. So, one time, the deadwood prostitute will offer John a freebie for being a regular customer and opens her legs very wide for John to see her virginal cavity. John is perplexed by the size of her clitoris. The prostitute had a protruding membrane so ugly John nearly threw up. He looked at the circumference of her outer virgina and senses a large circular membrane. He didn't know that the prostitute was fuming with stench and applied deodorant to cover it. The enlarged clitoris and outer part of the virgina were as a result of fungal invasion. The prostitute never knew orgasm since her puberty, at the age of thirteen, for she fucked ten men at that age.

John was delighted to see what he called a real pussy. He thought his wife was incomplete for she didn't have any of the prostitute's virgina quality. Amanda had a virgina sealed with the outer skin and John didn't see any of the interior part of her pubic parts. There was just a straight line of opening from the top to the bottom of the virgina. John never knew how the virgina looked until the sad prostitute spread her thighs so wide. John did his thing and departed. He didn't know that the odor of his penis had changed significantly. The prostitute mated with almost ten people a day and the strength of the odor of her virgina change by that number of times in a single day.

Lets say a woman begins as a virgin. When she mates with the first man to penetrate her virgina the scent of her inner part of the canal in her virgina resents the scent of the virgin man's outer skin of his penis head. The process results in the inner part of the virgina scent strengthening by two folds. At the same time, the scent of the virgin man's penis also increases by two fold. If the man mates again with another woman the scent of his penis head increases three times.

So also if the same woman mates for the second time with a different man the odor of her virgina is strengthened to three times. Hence, the prostitute that John fucked had the odor of her inner lining of her virgina change ten folds because she mated with ten different men. If the prostitute had sex with two thousand men before John fucked her then her inner lining of her virgina is a thousand times more stinky than Amanda. The odor of John's penis head also smells like the dirty prostitute and from then onward the cells of his head penis will revert and produce stench equivalent to that of the prostitute he fucked.

The next day, John desires to get blowjob from Amanda as he normally does. This time, however, Amanda nearly throws up from the pungent smell oozing from John's penis head. She asks John to wash his penis before she could do him. John realizes that he had taken shower an hour before and he resents Amanda immediately. He thinks she has slighted his manhood and demeaned him. And they result in a fight. Amanda never want to give John a blow John again because of the stench that oozes out of his rotten penis and that severs their cordial relationship. In no time John assaults Amanda and the police are involved. John is detained and sent to prison and Amanda desires to bail him out if he promised not to hit her again. That is why there is "zero-tolerance" law in Canada. Once a man is charged for domestic violence he is certain to be convicted and restrained from ever seeing his wife and almost all the time, his children. If you think this is a lie touch your wife with a twinkie finger and let her call the police in Canada.

Amanda desires to have John back because he is the father of her four children. However, the law says that he is a woman abuser. John merely pushed Amanda and she sat down on the couch with his hand force. He never knocked her head nor inflicted any pain on any part of her body. But Amanda knows the law like the palm of her hand and knows how to use it. She knew all along that her husband John was just a silly tenant in their home. She knew from the day she was married that she could take John to the cleaners anytime she desired. A landlord would give a tenant thirty days notice to vacate his premises but a fucked up vixen Canadian woman deals a blow on her lawful husband in a matter of one hour.

Now John is a fucked up white nigger and spends his time in prison full of rage and fucked his mind up. He spends his time as a lonely man and never realized that he was in prison for as much as two months. So, one day, the prison guard comes and gives him his release papers and gave his one-way bus ticket to Union Station in Toronto. John had no place to reside so he finds a Salvation Army shelter and remains there for the next two years. Meanwhile Amanda, his sickly, deadly Native Indian wife, has found a fucked up Cherokee clown who would fuck her ass like no one man has ever done. She desires to suck his dick he consents, she desires to lick his ass he consents, she sucks the scum from his penis and swallows it and he doesn't mind. So Amanda finds a man of her life. In the Berry Indian reserve she was just a tenant to any man, and a rib that she is, but in Toronto she became a landlord of her husband and a landlady because of the fucked up inferior culture of the demeaned white man.

John doesn't know that he is a saint who committed unforgivable sin that is judged by his maker only. He dealt no blow to his now impaired and sickly bitch that Amanda had become. He deletes his family memory by consuming alcohol and taking drugs. He finds his countrymen from nearby reserves with similar predicaments who were also inflicted by the messages of the inferior cultures of white farts. He spends his days totally lost and absolutely drunk and mangled memory from LSD and other deadly drugs. Then he suddenly realizes that he has a second chance. He pleads with Amanda to take him back for he missed his family terribly.

Amanda smells the stench oozing from his alcohol fumes emerging from the dying and once proud Assesena warrior of the Aboriginal Indian tribe. She lashes his with insults and condemns him as a worthless and useless man. John pleads for mercy and finds his heart bleeding with messages of divine. Amanda the dirty, however, knows the laws of the diddle head fag white man. She gives him a last chance and tells him to move in with his family feeling like a goddess knowing his entire life would be in hands. Soon, a message comes into her head and tells John to get out of her mind. John tells her she couldn't do that for the fine law of the fucked up white niggers doesn't give her that authority and get rid of him without a reason. So Amanda consults her evil mind. She call the police to evict John from his rightful home and never to see him again. However, the police officer tells her of the rights of John.

Amanda never gives up and would be determined to destroy John permanently. She knows when John would return from his alcohol binge and constant party. She slits her wrist and lies down in a water filled bathtub. Innocent John calls 911 not knowing the devil's evil intentions. The call doesn't get through so John tries again for the devil has disconnected it from its main jack. He doesn't know why the telephone was not dialing and tries to hook it to a different jack and finds out the reason. He reconnects the severed line and paramedics and police arrive instantly. John is relieved not knowing what tragedy is to come to him. The paramedics revive Amanda who would pretend initially that she was on the verge of death. The police ask Amanda why she attempted to commit suicide. Amanda makes a slurred statement accusing John of trying to murder her.

John is flabbergasted and couldn't believe what he was hearing. The white supremacist enforcers of the fucked up fucken white man mangle John to their lifeless trap cabin they call their cruiser. The fart judge sentences John to two years in prison for attempted murder. John imagines that he would get a quick release if he behaved appropriately in the Maplehurst Correctional Center, at Malton, at the outskirt of Toronto. However, he is told by the fine prison guards that he had to serve his full term because he was a second domestic family offender.

He finishes his prison term and can't wait to drink his first "spirit alcohol" for that is how they fucked up John's ancestors and stole the entire land of Canada by the faggot white human inferiors and rejects of their mind. John had no place to go when he was released from the Finch and Dufferin Native Indian court, a kangaroo court to appease John's family and destroy their memory of their ancestry. John finds Rachel Vance, a good fine but fucked up Jewish woman, a COTA worker. She finds John a residence at King and Dowling group home and John stays there for nearly a year. At last holy woman Rachel Vance finds John an apartment at Sherbourne government housing complex where John stayed for the remaining of his stay in the devil's city of Toronto.

John binged and abused LSD by panhandling at the intersection of Queen and Sherbourne. His head was never right and he visited his mother at the Assawagne prison tribe of the disoriented real owners of the vast land of God's homeland called Canada. He finds her deadly poisonous for she has been influenced by the lies of Amanda's family and the deadly prison record of the conviction inflicted by the fucked up and fart Dufferin and Finch court judge, witch judge Marshall. The fucken bitch judge Marshall warns John not to see his fucked wife and children to eternity. And that is how the proud warrior ends up as a lonely and desperate man.

His honorable mother at the reserve finds his actions on Amanda as a sickly mind. John doesn't know that his mother became his enemy by the devil's facts she assumed from the honorable family of Amanda, the saint but converted to a sickly and maligned mind, imposed on her by the deadly and inferior culture of the wife abusing evil tendencies of the deadened white supremacist, sadist, fart, immoral and dirty minds of the sickly white man. The white woman doesn't have the protection of a man and hence, imposed laws to protect herself from their evil tendencies and killing minds. The white woman is always described as a wicked and evil woman by her fellow white man and doesn't know how to protect herself except to destroy the mind, memory and honor of the fucked up goon. That is why Canadian law works on white fucked up maggots but destroys immigrants from much higher cultures including that of the saintly maroons, the Aboriginal owners of the land.

John becomes infuriated by his mother's serious words and strikes her with the booze bottle he carried with him to her head and she dies. He sits by her side not knowing what he did until the deadly white supremacist sheriffs are invited to conduct investigation. Two hours later John regains his memory, a little, from a combination of alcohol and LSD imposed hallucination and is told his dear and loving mother was dead. He cries bitterly and the fucked up sheriffs rule him out as a suspect and hence, John's mother death becomes a cold case file. The sinful and dirty sheriffs don't regard Aboriginal lives equivalent to that of a rat and hence close murder cases as soon as it opens if no immediate arrest is made.

So, John returns to the sin city of Toronto where the sickly and deadly white supremacist Bill Blair, the Toronto Chief of Police, resides. John doesn't know that the mind killer of immigrants Bill Blair knows the entire history of every immigrant Canadian. He regularly strips and deports, with alacrity, and revokes foreign-born Canadian citizenship at the slightest error in their application to emigrate to the devil's domain turned into nation of Canada.

That is why the citizenship certificate issued to foreign-born Canadians is less worthy that the toilet paper that you use to wipe out your shit. However, John is an Aboriginal Canadian and the devil Bill Blair could not have jurisdiction over him. Bill Blair couldn't have done it without the help of the sickly, devil and derailed mind of judge Julie Marshall. No other judge could do that and none of them did. Julie Marshall striped of over two hundred foreign-born Canadians for the slightest mistake in their original application to enter Canada, some dating to may decades. Death sentence was imposed on particularly Jamaicans even though no death sentence exists in the entire devil's nation turned country of Canada.

So John migrates back and turns into a mad man. He drinks excessively and hallucinates all the time from LSD and prescription drugs. Soon he realizes that he was victimized by the saintly and turned devil wife, Amanda. He finds her family name in the yellow pages and traces her address. By now she had sold her family home and was living in a government housing complex. She also was a booze binger even though she didn't hallucinate. John takes a bus to Jane and Sheppard, where Amanda resided and waits for her at the entrance of her home.

He delights in the bottle of booze in his hand and pops, every now and then, a pill to extend his hallucination. Amanda doesn't show up until late in the evening for her four children were now grown up and did the same thing their mother did. Jesse was a fucked up drunk at the age seven. Melissa was a tramp by the age of ten. Megan was no better than Melissa and popped up pills, was a regular hole by the age of twelve. Heather was the only survivor of the death sentence of the inferior white man's culture and that is where Amanda spent the day for she was cohabiting with a sickly, deadly, mind controlling, woman peddling pimp dickhead nigger. The sick nigger pimp doesn't even spare Amanda and could screw anything that has a hole. So John waits for her and eventually falls asleep at the entrance of her door.

Amanda sees the dilapidated John and accuses him of trespassing. John pleads with Amanda and begs her for forgiveness. She tells him that she was done with him and that he should never show up near her premises unless he preferred to go to prison again. John continues to plead and promises to guard her and his children with his life. He notices his visible slurring and fights him to leave. John didn't know that he had a penknife. When she struck on his thigh he noticed a sharp pain and knew it was that of the knife he used to open a beer can if the opening devise snapped. So he removed the knife from his pocket and stabs Amanda repeatedly on her thighs for he was kneeing to beg her. Amanda screams not loudly and no one comes out of the closely knit housing units of Jane and Sheppard. Amanda fights vigorously but John stabbed her relentlessly on the back of her head snapping her medulla oblongata. John didn't know what he was doing so he sits down and wipes his knife and smokes a cigarette.

Soon, he knocks on the next door of Amanda's neighbor and reports a homicide against his dear wife for he sobered after thirteen hours. He killed her at nine o'clock in the evening and reported her death at ten o'clock the next day. He had no clue how he got to Amanda's residence. So police rule him out as a suspect immediately for he was honest and believable in his testimony. And now John is back in the reserve of his Aboriginal tribe. He remarried and has two beautiful children growing far away from the inferior culture that evolved over the few decades and since women gained equality and organized themselves as a movement and enforced feminist ideology on the ignorant and feeble minded white men.

Man doesn't know that deadly women are as a result of his mind. Very few are women who invented anything in the entire history of mankind. Most of them are intellectually less beings and are created to be the subjects of man.

John is a preacher and has divine spirit in him. His tribal chief blew hot marijuana smoke through his nostrils for many times. He told him to smoke the holy pipe and John soared into heaven and returned to earth. His demented memory melted and desired to have new memory stored in his mind storage bank. Find him in the Berry reserve and confirm this story.

Hitler never desired to destroyed the Jews. They asked for it and he delivered unto them exactly what they wished for themselves. They stood firmly against Hitler and hence decided to exterminate them. Hitler wanted to become the German leader very badly and only the Jews stood on his way.

When the whole Germany rallied behind him the mangled mind of the Jews couldn't figure out his enormous popularity. Instead of joining the indigenous native Aryans they began to separate themselves which became convenient for Hitler to target them. All the German population knew where every Jew lived. They didn't know that there were so many Jews in their country. Hitler became enraged when he counted just how many Jews there were in his fatherland. There were six million Jews in entire Germany and all of them were concentrated in the major towns and cities of the country.

Hitler knew his grandmother was a Jew and didn't want to destroy them but they wouldn't stop demeaning him in their powerful newspapers. Hitler nationalized all Jews newspapers but they used other mass media to continuously try to fight him. Knowing that Hitler nationalized all mass media outlets and spread Nazi propaganda only. However, the Jews never relented. They continued to disseminate false information about Hitler.

Hitler industrialized Germany in just two years and the beggars Germans that they were, after the First World War, felt a new lease of life and earned wages to support their families. Hitler brought honor and dignity to the impoverished German nation. To the Jews, however, he was an evil mad man who perceived himself and other Germans equal to the sickly Jewish supremacists dogma that they are the only chosen people of God

Hitler knowing that invited Jewish leaders to strike a deal with them. He told them that he would appoint powerful Jews to his cabinet if they desired but that they should stop tarnishing his image. He told them that Himler was a good Jewish rabbi and others in his cabinet were prominent Jews. The death-wishing Jews rejected his offer and told him that they would demand nothing but his resignation because he had already tarnished their family names by equating Germans as a sacred people and the supreme race. Hitler didn't know that Jews saw themselves above other Germans. He instantly adopted a parliamentary edict declaring Jews as none Germans and stripped them of their citizenship.

This time the Jews regretted their stupid desires and tried to compromise with Hitler. Even the Catholic Church sided with the Jews and declared Hitler as the enemy of God. However, Hitler delivered deadly blow on the Catholic Church. He appointed his own cardinals and bishops who were Nazi sympathizers and the pope became insignificant in Germany for the leaders of the Catholic Church never totally communicated with him. This fact could be verified from Pope Benedict who is still alive and was a minor in Hitler's time.

So the Jews were doomed and they knew it. So many of them migrated to Palestine and many migrated to America. All of them would have evacuated Germany if a single Jew had not attempted to deliver death blow to Hitler.

Germany is a sickly white supremacist nation not because of Hitler. Hitler only knew of the deep thoughts of his people and didn't realize how much the Jews had destroyed the pride and honor of the white race in Germany. Hitler was not an idiot. He had a perfect mind-of-man and delivered his ultimate verdict on the Jews, mind manglers.

Germans are the original Aryan races and are indeed superior Europeans. The Jews, however, reduced them to zombies with their deadly mind controlling expertise for which they destroyed many societies in Europe before they migrated to Germany. Portugal deported them and it became a great nation. Spain deported them and they also became a great nation. Many European nations also did the same and they regained their freedom and liberties and became great nations.

Great Germany did not participate in the scramble for Africa much, the Americas and other continents because they were vassals of the Jewish, mind murderers. How can Portugal, Spain, France, England and even tinny Belgium acquire numerous colonies in the world when giant innovator Germany could have just one colony? That is how much Jews destroy the innovative mind of a people.

When the Ephraimites left the kingdom of Israel, in 731 B.C, they made their base in present day Germany and conquered the entire regions east of Germania (Germany), including present day Russia. Germania controlled the rest of Europe by what we know today as the great Roman Empire. All Romans, with no exceptions, were from Germania. That is why the European Union cannot bear any fruit without the full blessing of the Republic of Germany. Now they know who they are and they should treat their other European counterparts as brother and not as inferior Europeans, which present day

Germans think. All white people are genuine Israelites, descendants of the ten tribes of Israel. However, from Germany all the way to Russia are exclusively Ephraimites and the descendants of Ephraim, son of Joseph and son of Jacob, son of Isaac and son of Abraham our forefather. Africans are the descendants of exclusively Manasseh only and have the blood of Judah.

Unfortunately, today's Germans are the slaves of another brand of Jews. Hitler saved them from Jewish torment and degradation. Germans should not feel guilty for what Hitler did. Hitler saw his own manner of retribution and executed it brutally and with no mercy or sympathy.

The Ephraimites migrated further north and settled in present-day Germany because they were rejected by the other nine tribes of Israel because of their wild and controlling nature. The remaining nine tribes of Israel settled in Turkey for about a hundred years. The Ammorites were their neighbors. As the nine tribes of Israel farmed the Ammorites began to kill them every day. So, many of the nine tribes of Israel migrated further north in the manner of the Ephraimites but couldn't go eastwards because of the Ephraimites control of the entire Easter Europe, beginning from Germany. So they were forced to travel to westwards and are today the ancestors of the Western Europeans, including the British.

The remaining nine tribes of Israel in Turkey survived and eventually made a pact with the Ammorites and had peace for sometime. But the Ammorites never stopped harassing the remaining Israelites until they took arms to defend themselves. They dealt heavy blow on the Ammorites and restored peace and unity among them. These nine tribes of Israel would ultimately become the architects of the Byzantine Empire. Unfortunately, the Byzantine Empire would not last for more than one thousand years. The Ammorites, the ancestors of the present day Turks, would claim what they perceived as their motherland and exterminate the descendants of the nine tribes of Israel.

The Orthodox Church was the sacred messages of the nine tribes of Israel from which the Catholic Church eventually proliferated from. So the ruthless and murderous Turks are descendants of the Ammorites. They are certainly not Israelites. They are not the descendants of Abraham and don't share the blessing of God on earth. The earth is exclusively reserved for the children of Abraham and no other human being should live in it. The demented Muhammed slaves are deadly and terrorize the German people. They believe in stabbing and threaten to behead the fine people of Germany. Beheading is not a patent held by the evil Moslems only. Christians know how to use a knife too, and even better, know how to skin an infidel Moslems also. So, Israelite Germans are being threatened in their fatherland by infidels that are not even the descendants of the children of Abraham. And that should never happen.

Don't forget that Jews are not Israelites and are also not the children of Abraham. They are descendants of the Babylonian Empire which is not traced to our forefather Abraham and his descendants. When the Jews returned from Babylon they had only twenty percent Israeli blood. Over the years, the dominant eighty percent Babylonian blood in them had overwhelmed the twenty percent Israeli blood. Ar a result, Jews are wholly Babylonians with no trace of Israeli blood.

So, what is next? Next is the story of the nearly dead Americans and Canadian whites. No one knows the deadly Jews like the leader of the Nation of Islam, the Honorable Minister Louis Farakkhan. He blasted them at every occasion and warned mainly black Americans against the penetration and deadly effects of the Jews in the American society. White Americans never listen to his messages and they lost their precious minds as a result.

How did the Jews become white-man mind killers, especially in America? When Hitler exterminated many of them they made their journey to the United States of America. American president Theodore Roosevelt didn't want them to invade his country. He returned many ships bound for the United States of America. But the ships continued coming. At the end Roosevelt relented and allowed them to stay temporarily mainly in New York and its environs. Roosevelt didn't want to give them American citizenship so the Jews devised to kill him. A Jew volunteered to delete Roosevelt. He deleted Roosevelt when he poisoned his drink in a derby that Roosevelt attended. Roosevelt collapsed but he didn't die but his legs were permanently paralyzed from the poison for his heart stopped for two minutes.

So the Jews thought of a different way of assassinating Roosevelt. They made a pact among themselves to spread false information about the conducts of Roosevelt. Roosevelt didn't know why the American people were spreading false rumors about him. So he didn't respond and waited for the lies to die down. But they kept on coming and the Jews formulated more and more false stories until Roosevelt's reputation was totally destroyed beyond redemption. Roosevelt couldn't dispute these facts but still didn't respond. After that the FBI realized that the newly arrived Jewish immigrants were responsible for the falsehood So, Roosevelt regained his reputation and even though he became a very weak president after compromising with the Jews who restored his damaged reputation and stopped their propaganda against him

How could the Jewish immigrants attain such an overwhelming power to the extent of ruining the reputation of an American president barely five years when they entered the country? The answer is the evangelists. They told the evangelist that they were the children of Israel and "the only blessed chosen of God" and that damnation will reign on the United States of America if they deported them. The entire evangelist movement became the main instruments and weapons of the Jews to destroy the white race. Whatever the Jewish rabbis told the evangelists they repeated it accurately as if it was written in the gospel of Jesus Christ. That is how the evangelist morons destroyed the United States of America and destroyed the minds of the white race.

Why did the evangelists fall for the sinister tricks of the Jews? How did they manipulate their minds to screw them so badly to the extent of reducing the white race to second class level? And the answer is in Paul's letters in the New Testament. Paul was a Jewish supremacist and delighted in his Jewish origin. He wrote in his letter that salvation was to the "Jew first and then to the Gentiles." That sentence screwed up the entire Christian faith and distorted the sacred messages of the New Testament and the four gospels. How could salvation come to the Jews when they deleted the Son of God and crucified him? That way Paul was able to obliterate the perfect message of the four gospels and distort them severely.

When Jesus was alive he saw the Jews as enemies of God and they had never been believers in the Lord even since the time they were sacked and exiled into Babylon. Deadly Jews have been a threat to the Lord since that time. Jews believe in the Ten Commandments but their rabbis have multiplied them to more than three hundred rules. The Torah and its Jewish laws is the constitution and the omerta of the Jewish cult. Heavy penalties are imposed on Jews who break any of the more than three hundred orders. That is how the Jews operate and that is why they are always successful. Their absolute unity is their strength. They all think alike.

When the Jews entered America they didn't come from Africa or other continents. They mostly came from liberated Germany. In Germany they had mind-controlled the fine citizens of the Aryan nation for more than four hundred years. There, they succeeded by demeaning the sad German citizens by falsehood and fabrications to destroy reputable Germans who did not tow their lines. For example, Otto Bismarck, the German Chancellor, just a little while before the ascent to power of Adolph Hitler wanted to single out Jews who were detrimental to the German economy. The Jews realized that and they desired to falsely destroy him. However, they were not successful initially but evertually succeeded.

Manna will never fall on the United States of America until they clean their nation out of the Jewish fallacy and return to the blessings of the Son of God, Jesus Christ.

When the dirty mannered Jews finally deleted Roosevelt they realized that their destiny was in their hand. They instantly knew that they could conquer and dominate America very easily. And it didn't take them much time to do so. Mind you, the Jews that migrated, nearly all of them, came from Germany. There, they understood that Hitler was too popular for them to control him. Hence, they ensured that no American was too powerful and popular to become a threat to the Jewish nation of the occupied land of Palestine and to the entire Jewish population throughout the world.

The Jews initially made the continent of America their base and became a nightmare to the silly people of Africa. Africans who deviated from the directives of their secret service, the Mossad, became instant targets and eliminated them. Kwame Nkrumah became their deadly enemy because he was advocating for a unified army to protect every country in Africa instead of individually having a very expensive army. The Jews did not want to see the continent achieve that and hence, overthrew him and replaced him with Bussia, a timid and submissive Ghanian. Africa lost probably the best leader in its entire history after the colonial period when Nkrumah was removed from the helm of power. Nkrumah was subsequently murdered in Romania where he sought asylum.

So many African dictators were ruled from Tel Aviv and were under the direct influence of the Israeli prime minister. Those who deviated slightly were eliminated instantly. No

African leader dare disobey the instructions of Natanyahu until date. Natanyahu is the life-giver and life-taker of the entire continent of Africa. Just ask Iddi Amin Dada. The poor soul was an Israeli stooge until the Palestinians hijacked a plane and landed it in Kampala, the Ugandan capital. He made an error to desire the Israelis to negotiate for the release of the plane and its passengers.

That was an insult to the death spewing maddened Jews. So the Israelis took everything into their hands and did a very daring rescue of the hijacked plane and its passengers. In the process, eight commandoes from the death-squad who murder innocent Africans were deleted. That was enough to permanently destroy Iddi Amin Dada. Amin knew how deadly the Jews were as a result, he started killing many of his compatriots that were suspected of being Mossad agents. But the Jews were not satisfied to see Iddi Amin alive. So, they began to spread false and fabricated information to destroy Amin. The devout Moslem was accused of eating human flesh and described as a cannibal and the world bought their lies. Eventually, Iddi Amin was overthrown and took sanctuary in Saudi Arabia. The Jews have excelled, almost to perfection, in the spread of utter falsehood and to destroy any of theose who stood against them and asserted themselves as genuine Israelites, as Africans and Europeans are. The Jews are merely Babylonians.

The Bible says that Jerusalem shall be "trampled upon by the Gentiles" for forty-two years and that is what happened when the European Jews, Gentiles, conquered Jerusalem in 1967 and ruled it for forty-two years when their cover was exposed and identified as non-Israelites, as I am doing today. No sane person will know them as Israelites again and they shall be relegated as liars, deceptors, fiction aspirers and utterly tale fabricators in the future. Forty-two years ago, to be exact, the Jews declared Jerusalem as their possession and now claim it as their capital city. This is a fulfillment of Bible prophecy. It is written that the Gentiles would trample upon Jerusalem for forty-two years before the end of time comes.

The Jews killed Muritalla Mohammed, perhaps the greatest leader of Nigeria. He was said to have been killed by a renegade soldier named Dimka who was instigated by the Jewish state.

The Hausas are nomadic tribes that originated in Eritrea. About five hundred of them from the town of Senafe migrated southwards and settled in a place called Intoto, present day Addis Ababa, in January 10, 1075 AD. They named themselves Hausa because they were a union of several clans gathered from the same tribe of Akeleguzay in Eritrea. The word Hausa means "Union" in Tigre, which is the original language of Eritrea. The Hausa were constantly threatened by all their surrounding Lamanites. The Hausas were predominantly Nephites and much fairer in complexion than the black Lamanites, Amharas of Deqi Lema.

There was no water in Hausaland, present day Addis Ababa, and had to dig so far away to find well-waters. As a result they depended on pond waters because of the abundant rainfall in the region. However, during the short dry season they almost died of thirst and had to ration the little water they obtained from the few wells they had. For four centuries the Hausas survived but did not multiply greatly because of the hardship they experiences, especially the lack of abundant water resources. Their original number was 10,000 and four centuries later they barely multiplied and remained at 13,000. Bilharzia was the main killer of the Hausas.

Bilharzia is not a disease and is not caused by any virus. The Hausa ate the same vegetables their herd ate. The Hausas never ate food that their, especially goats, sniffed and rejected as food. So, when they devoured a certain strange plant they immediately adopted it as their main staple food. The plant was teff, now consumed predominantly by the Amhara tribe in Ethiopia. However, too much consumption of teff was killing them in large numbers and would realize it much later after they migrated away from Intoto.

Too much consumption of teff, especially white teff, produced too much iron in their blood system which clogged their arteries and they died of heart failures. Amharas have developed mechanism to adjust the amount of iron in their blood system over the centuries, however, if any strangers consume the said teff for a few days they feel choked when they consume it initially and that is the body's way of telling the consumer that he is eating a very toxic substance. Everyone that feels choked when consuming teff should never eat the plant ever unless they want to experience an untimely death and a massive heart attack at the end.

So the Hausas began to lament and pray to the Lord to give them a better land and away from the enemies they were surrounded with. The Showaian Amharas in particular constantly harassed them and deleted them sometimes when they farmed in their land they perceived to be their own. At that time the super power of the region were the Mensas.

In 1358 AD the Bejas were ruled by a devil queen named Yodit. Yodit was a sickly and deadly woman. She is known for her cruelty all over present-day Ethiopia, Eritrea and Northern Nigeria. Yodit was never married and never had her own children. She slept with so many men but could not produce an heir to her rulership. She became a very disgruntled woman. She suspected that she became infertile because of the syphilis infested Amhara handsome prince she was anointed, by her father, to marry. The said man died shortly after her wedding to him and knew that he had puss all over his penis head. She didn't know that she carried the death causing bacteria in her blood stream.

So, immediately her ruler father found her a Bejan nobleman, for that is what Bejans did to their widows, but the second husband died shortly after with the same disease that her first died from. So the Bejan ruler sought another suitor for his daughter and that went on until Yodit killed seven men she was married to with the same disease. The Bejans mentioned their concerns to the kind and told him that his daughter was a Satan and no Bejan would marry her unless the person wanted to commit suicide.

But the ruler did not relent and secretly sent her men to mate with and all of them died shortly as the others and Yodit never got pregnant. So, Yodit realized that she was born to be the ruler and killer of men and told her father so. The Bejan ruler was very amazed but he didn't think that his beloved daughter would murder him in his bed. Hence, Yodit anointed herself as the queen of the proud Bejans and ruled them with ruthlessness. She mated mainly with many Amharas and after having sex with her she ordered their extermination thereafter. No Bejan was ever killed. Initially, when she was young all her husbands were Amhara princes. So she continued mating with them exclusively for she resented the Bejas even though she was their ruler.

So Queen Yodit began to seek new lands to expand her territory for Gondar was a very small area for such a power-thirst queen. So she inquired from Bejan merchants and they told her that there was so much land that reached the end of the world. She began to be fascinated by such adventures to occupy all the land to the end of the world. She told the Bejans if they would accompany her to reach the end of the world but all her local members of her cabinet told her it was not a viable investment. But she wouldn't relent.

Then she discovered there were a group of people who prayed God to give them a better land. So she consulted with the Hausas and they were all excited because they were honored by the great queen. But they didn't know that Yodit was a ruthless woman that murdered all men that she mated with. She killed thirty to forty people every year. So the Hausas packed all their belongings, herds, children and all members of their families and followed Queen Yodit blindly. The Hausas were not many but numbered 13,000 only. The colonialists travelled across the Sudan and didn't settle because there was no sufficient vegetation for their cattle. But regardless, Yodit ordered a few Hausa to settle there to ascertain her domination of the Sudan. She went further and came to a river that was muddy and called it Mai-dugri (Huskey water in Tigre) by those Hausas who settled in the region.

Yodit had ten thousand soldiers of her own apart from the Hausa population. So the war went on for two full days for the Yaredites attacked them even in the middle of the night. At the end nearly all the Yaredites were deleted and the rest were scattered all the way to the south of Nigeria. So, the Hausas called the place Zeria for it means "she scattered." Note this important fact. Zeria means "she scattered" and not "he scattered" because Yodit scattered them. The gender is very relevant. Present-day Hausas call Queen Yodit as Queen Amina and the history narrated by Maitama Sule is nearly as accurate as the true history of the Hausa people.

So, Yodit proceeded further into West Africa leaving many Hausas in many convenient location to guarantee her control of the earth until the end of the world. Eventually, she reached the Gambia and knew that she could not proceed any further and knew that it was the end of the world. So she began to trace her footsteps and to return to her motherland. She travelled eastwards in the same path she came. Aware of her map she realized that those she left in certain location were not inhabiting the land she conquered as her own. She wondered why. She saw nearly barren land with no inhabitants but noticed well organized houses with burning fire in their kitchens.

So she knew they were around but probably gone for hunting for she couldn't understand why even children and women would go hunting with their elders. So she searched the nearby forest and found all the albino inhabitants of the village clamped together with terror. She asked them why they were scared. They told her because they had killed the dark people she left behind in their village. So she asked them to return to their village to render her judgment. They did so by all walking in a single file touching each other's shoulders. Yodit asked them why they were doing so and the Tigre interpreter told her that Albinos don't see well in bright sunlight. So she pulled them all to death and left others behind to settle on their land instead

So she continued her journey back to her motherland and saw the same thing. New homes with no inhabitants. She searched the nearby forests and found the Albinos. She murdered them and left some behind from her Hausa entourage and would become the ancestors of the Fulanis. She would do the same thing all the way back to the Sudan. There she fell ill and died in Kirtim ("dead" in Tigre) which is present-day Khartoum. The Hausa would become very powerful and control very vast lands in many parts of West Africa. They will produce many powerful leaders who will rule them with utmost wisdom just like the Kentibas in their original fatherland of the holy land of Mensa.

The Waziri of Maiduguri, the Waziri of Sokoto, the Waziri of Katsina, the Waziri of Keno and other smaller kingdoms would produce great minds and scholars. The Tuaregs will also become great scholars of Timbuktu and research their minds just like the Mensas.

These Arnes are still in Mauritania, Morocco, Libya, Algeria, and Egypt where only the Coptic Christians are Menassehites. The Tunisians are the only Syrians who migrated to the sacred land of Abyssinia (Africa) in the year 1599 AD. These Arnes, Arabs occupying Algeria and some in North Africa, are illegal immigrants and should be seen as such for they eliminated the black race to settle in those regions. All Mensas should gang against them and throw them into the sea without any mercy for they are the arch enemies of the Abyssinians who are the blessed sons of Abraham, Isaac, Jacob and Manasseh. Their extinction is imminent and it is just a matter of time.

"You bastard arnes. You don't have the precious blood of the Mensas. If you ever mention the name of my fatherland in vain again I will wipe out all of you," said the angry Zerian. But the Madiba was adamant and told the Waziri that the Biroms would rather die than give up the name of the land of their ancestry. Then the Waziri said, I know why they called you Bihirom. You became evil and that is why the Mensas kicked you out of your fatherland. The Hausas left voluntarily from Senafe and they are the only Abyssinians with the same pride of the original Mensas.

All of you people throughout the entire continent of Abyssinia were thrown out of your fatherland of Mensa because you committed one crime or another against the people of the Lord. All of you have one curse or another except the Hausa. Now I know who I am and I know all of you are my brothers but are inferior to me. If you devil worshipping bastards call the name of Waziri in vain you die a miserable death. Forget about calling your leaders Waziris that will never happen because you will never ever have a wise Mensa leader until the end of time, like my father and grandfather and great grandfather etc.

So the Waziri looked around and saw a variety of people gathered surrounding him. He pointed his finger at the one closest to him and said, whose cursed name do you bear? The man bowed his head and said, I am John Igba, from the tribe of Igala. The Waziri figured out what his tribal name meant and said, shame on you Igala. Your mother searched the Kentiba's head and fucked him up that is why the Mensas cursed you and named you Igala.

The most honorable Waziri, I know the meaning of my name and also know the cursed names of all these that you call your subjects, said the leopard face Igala. The Waziri was astonished and said, Hausas are the only people who spoke Tigre, our mother tongue, I despise you for dishonoring the wild and distinguished mind of the Waziri, the "My guide of God." O!

The Waziri looked intensely at the fine tribal marks of the Igala man and said, I know who you are and your ancestry as a Mensa. The Igala man replied, I know what Igala means "Her son" and that I am a Mensa by origin. So the Waziri looked at the Igala man and saw a deadly rival. He said, how many wives do you have Igala man? The Igala man said, I have just two, one for breakfast and one for lunch. The Waziri said, why don't you have a third wife for dinner if that is why you married them for? I don't know why you want me to have three wives when you have just one, replied the arrogant Igala man.

Igala man pointed to the muscular man next to him and said, this black man sitting next to me calls his tribe Igbo. Igbo means, "Submission" in our language of Tigre. This messenger is my brother and calls himself Idoma. Idoma means "Prohibition" in our fine and beautiful language of Tigre the source of happiness and poets throughout the Adam land.

Oh great Wazir, I am an Ibibio and I know the meaning of my tribal name. Ibi-bio (Ibbi biiyo) means "Shenanigan-by works." That is why I am the most hardworking member of Hausa land. Another man rose up and said, I am Oduduwa's son and my ancestor landed on Olumo rock and called himself Yoruba man. Odu-diuwa means "His tree-medicine" and I am an expert in herbal medicine throughout Abyssinia land. Olumo is the sacred mountain of my ancestor and Elimo means "Education" and that is why Olumo people are brilliant a little less than this Igbo man. Oruba means "Departure.".

Another man got up from his seat and said, I am a Tiv but my fatherland called T'iw because I snapped goat's neck with my bare hand. I know that I am the most ruthless man in your Hausa empire. I snap my opponent's necks just like my ancestral name for Tiw means "Twist." I beg to differ your honor. I know the Tiv is a ruthless man and I fear and respect him greatly but I am a Jukun man. And I know that I am the most ruthless and wicked more than any man in the universe. Jukun (chukun) means "Mass-murderer."

I know you are sad great Serki of Hausa land. I am the Oba of Benin and delight in my shenanigan that is why I established the kingdom of Benin. For reason I don't understand these that surround my homeland don't respect me. Now I understand for this Igala man has opened my eyes and I am originally from the saintly people of Mensa land. Benin means "Evaporator" and all my pride and highness is just a sham.

I know who I am now said another one. I am Angas and all I want to say is that Angas means "Crowner" and if I don't accept the Serki of my mother I behead him with my bare hand.

The Waziri didn't know how to stop them but it is his duty to listen attentively to his subjects. Another man got up and said I am Shekiri, the only one who shakes when he sees his Waziri and that is why I am his favorite. Shekiri means "My beloved one."

The Waziri was appeased for a while until another one rose and shook his head and said, I am Effig and don't delight in the Waziris sexual prowess. Effig means "Sexual deviant."

Another one also rose and said, I know my own name, my family name is Igbira and Igib-ra means "Faith of vision." Knowing that the Waziri was alarmed and didn't know that other tribes saw visions like his forefathers, the Mensas.

The Waziri just bowed his head and said, I am the forsaken mind-of-man. But the Igala man kept on coming and coming with all the tribal names in the continent of Africa and knew why they were all driven out of their father and motherland of Mensa land. But the Igala man wouldn't forget the sickly Mensa ancestors of some fine tribes of the numerous Hausa land.

I beg to differ shouted the arrogant Igbo man. I call my leader Eze and the name means "Prosperity" in my forefather and foremother of tribe of Mensa.

All African tribes migrated from the Mensas. History would show that all Africans and the entire black race are brothers and sisters. The same analysis would show that Europeans are descendants of the gentiled (separated) ten tribes of Israel and are brothers and sisters also. Africans are the descendants of Mannaseh and the tribe of Judah only as could be verified from the African Bible, the Book of Mormon.

Ultimately, all the children of Israel, the Africans and Europeans, shall know their true identities. The Jews see themselves as the exclusive children of Israel and the "only chosen people of God" and have anointed themselves as the only authentic Israelites.

When the Jews returned fro Babylon and settled in in the Kingdom of Judah, they had only twenty percent of Israeli blood. Over the years they have lost that percentage of Israeli blood and the dominant Babylonian gene has overwhelmed the twenty percent Israeli gene and today Jews are pure Babylonians and share no Israeli gene in their blood.

Whites and blacks shall untimately come together and identify themselves as Israelites and form a perfect harmony witrh each other and shall prosper and live together as a single family members. The perceptions and demeaning distortions that had reduced, especially the white people, to the subservience of the Jews must be revisited. The evangelists propagate their falsified propaganda that Jews only are the authentic children of Israel and the only chosen people of God.

The Jews had used their Israeli status to gather themselves and unite in their perception and their ambitions to trample upon anyone who dared question their belief and aspirations as the only chosen people of God. No one is spared who speaks against them. The white people have specially suffered the brunts of these misconceptions and today whites believe that

God only favours the Jews exclusively. They see themselves as second rates in the eyes of God, next to the Jews, which is not true.

Africa had suffered tremendously from the wrath of the Jews also. As long as the state of Israel exists, the continent will never prosper without interference. African leaders would have to show ultimate loyalty to Tel Aviv if they are to survive. Those who are Israeli stooges are protected and groomed even if they are declared dictators in the continent. Ethiopia is highly respected by the Jews and Israel and their hatred and animosity towards Eritreans is very apparent. Israel fully supports Ethiopia against its genocide against Eritrea and the Eritrean people.

In America, many have become Jewish stooges. If an American elected representative of the people is to survive he has to tow the lines of the Jews and has to express unflinching support to the state of Israel. No American can aspire and grown if he dares speak against the state of Israel.

The Jews have successfully colonized America, as they did to Germany before, and from America they have extended their tentacles and reduced other nations to rubbles. America has become the enemy of the Moslem world because of its unflinching support for the state of Israel. Even if many Americans know the pathetic state of their nation they don't have the gut to confront the Jews and speak the ultimate truth that the state of Israel is built on the occupied land of the Palestinians.

Most damaging aspect of the Jews is on psychie of the white race in America. The once proud and arrogant whites who saw themselves as the blessed children of God when they came to America are today made to blieve that they are the second-rate children of God, after the Jews. The Jews have successfully imparted their grand perceptions and destroyed the minds, pride and aspirations of mainly whites in America. White Americans are today singers of the gospel according to the Jews. They willfully misqoute the Bible to justify that the Jews are superior to them falsely. The fear of the Jews in America is immense and they wouldn't dare utter any words agaisnt the Jews even in their dreams.

America is manily the haven for Jews primarily and the American dream is mainly alive for the Jews primarily who are the most prosperous among all the other ethnic groups in America. American elites will continue to dwindle in faith and moral decay until they assert the truth that they are the chosen people of God and that they are not second-rate choices in the eyes of the Almighty God, as they are today reduced to by the Jews. The whites and blacks, the true Israelites, must ward off the perceptions of the Jews and indemnify their status as the chosen people of God and His Blessed Son, Jesus Christ.

The above message is bound to resonate easily among blacks and Europeans who had liberated themselves from the Jews, as a result of Hitler's genocide, but the panic striken American whites will find it frightful to voice it out for fear of damnation from the Lord, a notion that was inserted in their minds by the Jews. White Americans have been destroyed drastically and see themselves as second class citizens even in their American nation, immediately after the Jews. They may rule America but deep in their minds that they could only succeed if they kiss the butts of the Jews and propagate primarily Jewish interests and uphold and pray for the prtotection of the state of Israel.

In recent days, Natanyahu, the Prime Minister of the state of Israel, had weilded the control he has over America and denegrated President Obama, as if he was his house boy, and bypassed him to address the Congress who are traumatized by the excessive powers of the Jews in America. Obama refused to meet with Natanyahu, being non-demeaned American, as a black man, but the white Americans were seen clapping and grooming Natanyahu, shamelessly, in the House of the American people, in Capitol Hill.

Africa shall never rise until the untruthfulness of the Jews is exposed and that is what I am doing. Blacks and whites are Israelites and they shall now know their true identities and shall rise together and establish perfect harmony between themselves when the catalyst of their disparity is eliminated. No one shall see the Jews as even Israelites lest the chosen people the Lord. Blacks and whites are the chosen of God and authentic Israelites. Period!!!

The Christian faith is fading fast because of the perception propagated by the Jews and who have proclaimed themselves as the only chosen people of God. Sadly, many evangelists have bought into this fallacy and have become the instruments of the Jews who would ruin them if they did not tow the line they have established and how the gospels should be preached and how Christianity should be observed.

If Jesus was crucified by the Jews, as the Bible indicates, how could the Lord proclaim the murderers of his Son, Jesus Christ, as the only chosen people of God? That is not plausible unless the Almighty God is a fool. The fact is that Jesus came to be the "light unto the Gentiles." The term "Gentile" means "Separatists" in Tigre, the language of the original Israelitres and the language of God, and is a reference to the separatist ten tribes of Israel who eventually migrated to Europe and are the forefathers of the white race.

The error many Chriatians make is quoting Paul who claims that the gospel came first for the Jews and then for the Gentiles. And that has become the theme of Jewish Messianics to promote themselves as the most superior human beings of the world and in the eyes of the Lord. Nothing can be further from the truth! Jews are not Israelites and Jesus never came for the Jews for there is no such indication in the four gospels. The Jews can continue to deceive themselves and anoint themselves as the most chosen people of God, but no Christian should fall for such absurd proclamation.

Israelites, blacks and whites, are the most chosen people of God. Whites are the ten tribes of Israel and blacks are of the tribe of Manasseh and that could be proven if the Bible and the Book of Mormon are interpreted accurately. Paul was a Jewish supremacist and his writings should be read with caution. Very few evangelists quote the four gospels today and indulge in promoting Paul's doctrine instead and that is not right.

The Jews shall continue to exist and they shall go on believing their old dogma, ever from before the advent of Jesus Christ, and they shall propagate their erroneous perceptions to eternity. But we don't have to believe them and allow them to interfere in our Christian faith and forcibly, and with intimidations, sway Christianity to their personal benefits and to acclaim their self-anointed propaganda that they are the only chosen people of the Lord.

Nowhere in the Bible that says that the Jews will be gathered someday and return to Israel, as the Jews insinuate and imbecile evangelists propagate. The Bible says that "Israel" shall be gathered from all the four corners of the earth. The Jews don't represent "Israel" and don't even have Israeli blood. The Europeans are the ten tribes of Israel and the black race are the descendants of Manasseh, son of Joseph, grandson of our forefather Abraham. At the end of time, which is now, the Lord will gather all Israel, that is, blacks and whites, and let them know that they are the authentic children of Israel.

Today, the whole world is deceived by the Jews and proclaiming them as the only children of Israel and the "only chosen people of God" which is totally untrue. Even from the beginning, only Judah was taken to Babylon and kept captive for almost a hundred years and not the northern Kingdom of Israel, which had already migrated to Europe. Hence, the Jews could not claim to be Israelites and the representatives of the whole of Israel. If they had made claim as descendants of Judah one would have considered their bogus claims. But to claim themselves as descendants of the whole of Israel is a blasphemy and utter and absolute fabrication and totally untrue.

The white race, particularly in North America, have totally succumbed to unfounded lies and fabrications. Everywhere in the American and Canadian churches the faithful Christians have bought into fallacy that Jews are the only chosen people of God and declared themselves as second choices, in the eyes of God, proclaiming the Jews as the primary and only chosen people of the Lord. The whites use to see themselves as the chosen people of God prior to the advent of the Jews, during the Second World War. Since then, however, their core belief has been altered and have been forced by the Jews to relegate themselves as the second choice of the Lord, the Jews being the first.

Whites and blacks and the other races enter the kingdom of God after the Jews have been accepted by the Good Lord, according to the fictitious propaganda preached today. That is what the fallacy that is engrained in the minds of particularly white people in North America and the white world. Hence, the white people are gradually abandoning the Christian faith and rejecting the Lord, whom they assume only cares for Jews. That is how much the the fabrication has demeaned the white race in particular. Luckily, not many blacks have fallen for such distortions and hence, more blacks are seen to worship the Lord and see themselves as the chosen people of God, with the exceptions of the few money clamouring black evangelists who are the mouthpiece of the Paulinean doctrine.

The Lord told Abraham, our forefather, that the Good Lord will bless those who bless him and shall curse those who curse him. The Jews have claimed this statement in the Bible and used it to their advantage and made us believe that anyone who spoke against the Jews will be damned and cursed by the Lord. That is absolutely untrue. Whites and blacks are the true Israelites and anyone that desires evil against us will be cursed by the Lord. That is what the verse says.

Today, Christians preach the so-called Judeo-Christian doctrine which is the gospel propagated according to the Jews. The Jews have become the sole interpreters of the Bible, even the New Testament. Even if they absolutely reject Jesus as the Son of God, Christians have given them the authority to dictate terms upon them with regard to how the Christiian faith should be preached.

Even the Catholic Church and other prominent Christian churches preach such fallacies today. The Christian world has become the mockery of the Moslem world and see Christians as betrayers of their faith by proclaiming those who crucified Jesus Christ as the only chosen people of God. If such a distortion is the truth who would want to worship the God that is that stupid to anoint the killers of His Blessed Son, Jesus Christ, as his ultimate and only chosen people to enter the kingdom of God, as the Jews make us to believe.

The Bible says that Jerusalem shall be trampled upon by the Gentiles for forty-two years before the end of time comes. which is now. And that prophecy has been fulfilled for the Jews, Babylonian Gentiles, had trampled upon Jerusalem since 1967, when they conquered the city from the hands of the Palestinians and eventually made it their capital city.

If you want to know how whites are the authentic children of the Gentile ten tribes of Israel, read my book, "The Revelations from God." If you want to read how Africans are the descendants of Manasseh read my book, "The Africa Bible" found on amazon.com

Printed in the United States
By Bookmasters